Lecture Notes in Computer Science 2594

Edited by G. Goos, J. Hartmanis, and J. van Leeuwen

T0226089

Springer
Berlin
Heidelberg
New York
Barcelona
Hong Kong
London
Milan
Paris
Tokyo

Andrea Asperti   Bruno Buchberger
James H. Davenport (Eds.)

Mathematical Knowledge Management

Second International Conference, MKM 2003
Bertinoro, Italy, February 16-18, 2003
Proceedings

Springer

Series Editors

Gerhard Goos, Karlsruhe University, Germany
Juris Hartmanis, Cornell University, NY, USA
Jan van Leeuwen, Utrecht University, The Netherlands

Volume Editors

Andrea Asperti
University of Bologna, Department of Computer Science
Mura Anteo Zamboni, 7, 40127 Bologna, Italy
E-mail: asperti@cs.unibo.it

Bruno Buchberger
Johannes Kepler University
Research Institute for Symbolic Computation
4232 Castle of Hagenberg, Austria
E-mail: Buchberger@RISC.Uni-Linz.ac.at

James Harold Davenport
University of Bath
Departments of Mathematical Sciences and Computer Science
Bath BA2 7AY, England
E-mail: J.H.Davenport@bath.ac.uk

Cataloging-in-Publication Data applied for

Bibliographic information published by Die Deutsche Bibliothek
Die Deutsche Bibliothek lists this publication in the Deutsche Nationalbibliografie;
detailed bibliographic data is available in the Internet at <http://dnb.ddb.de>.

CR Subject Classification (1998): H.3, I.2, H.2.8, F.4.1, H.4, C.2.4, G.4, I.1

ISSN 0302-9743
ISBN 3-540-00568-4 Springer-Verlag Berlin Heidelberg New York

Springer-Verlag Berlin Heidelberg New York
a member of BertelsmannSpringer Science+Business Media GmbH

http://www.springer.de

© Springer-Verlag Berlin Heidelberg 2003
Printed in Germany

Typesetting: Camera-ready by author, data conversion by PTP-Berlin S. Sossna e.K.
Printed on acid-free paper   SPIN 10872514   06/3142   5 4 3 2 1 0

Preface

This volume contains the proceedings of the *Second International Conference on Mathematical Knowledge Management* (MKM 2003), held 16–18 February 2003 in Bertinoro, Italy.

Mathematical Knowledge Management is an exciting new field at the intersection between mathematics and computer science. We need efficient, new techniques, based on sophisticated formal mathematics and software technology, to exploit the enormous knowledge available in current mathematical sources and to organize mathematical knowledge in a new way. On the other side, due to its very nature, the realm of mathematical information looks like the best candidate for testing innovative theoretical and technological solutions for content-based systems, interoperability, management of machine-understandable information, and the Semantic Web.

The organizers are grateful to Dana Scott and Massimo Marchiori for agreeing to give invited talks at MKM 2003.

November 2002

Andrea Asperti
Bruno Buchberger
James Davenport

Conference Organization

Andrea Asperti (Program Chair)
Luca Padovani (Organizing Chair)

Program Commitee

A. Asperti (Bologna)
B. Buchberger (RISC Linz)
J. Caldwell (Wyoming)
O. Caprotti (RISC Linz)
J. Davenport (Bath)
W. M. Farmer (McMaster Univ.)
H. Geuvers (Nijmegen)
T. Hardin (Paris 6)
M. Hazewinkel (CWI Amsterdam)
M. Kohlhase (CMU)

P. D. F. Ion (Michigan)
Z. Luo (Durham)
R. Nederpelt (Eindhoven)
M. Sofroniou (Wolfram Research Inc.)
N. Soiffer (Wolfram Research Inc.)
M. Suzuki (Kyushu)
N. Takayama (Kobe)
A. Trybulec (Bialystok)
S. M. Watt (UWO)
B. Wegner (Berlin)

Invited Speakers

Massimo Marchiori (W3C, University of Venezia)
Dana Scott (CMU)

Additional Referees

G. Bancerek
P. Callaghan
D. Doligez

R. Gamboa
B. Han

G. Jojgov
V. Prevosto

Table of Contents

Regular Contributions

Digitisation, Representation, and Formalisation
(Digital Libraries of Mathematics) 1
   Andrew A. Adams

MKM from Book to Computer: A Case Study 17
   James H. Davenport

From Proof-Assistants to Distributed Libraries of Mathematics:
Tips and Pitfalls 30
   Claudio Sacerdoti Coen

Managing Digital Mathematical Discourse.......................... 45
   Jonathan Borwein, Terry Stanway

NAG Library Documentation 56
   David Carlisle, Mike Dewar

On the Roles of LaTeX and MathML in Encoding and Processing
Mathematical Expressions... 66
   Luca Padovani

Problems and Solutions for Markup for Mathematical Examples
and Exercises 80
   Georgi Goguadze, Erica Melis, Carsten Ullrich, Paul Cairns

An Annotated Corpus and a Grammar Model of Theorem Description ... 93
   Yusuke Baba, Masakazu Suzuki

A Query Language for a Metadata Framework about
Mathematical Resources 105
   Ferruccio Guidi, Irene Schena

Information Retrieval in MML 119
   Grzegorz Bancerek, Piotr Rudnicki

An Expert System for the Flexible Processing of XML–Based
Mathematical Knowledge in a PROLOG–Environment 133
   Bernd D. Heumesser, Dietmar A. Seipel, Ulrich Güntzer

Towards Collaborative Content Management and Version Control for
Structured Mathematical Knowledge 147
   Michael Kohlhase, Romeo Anghelache

On the Integrity of a Repository of Formalized Mathematics 162
    Piotr Rudnicki, Andrzej Trybulec

A Theoretical Analysis of Hierarchical Proofs 175
    Paul Cairns, Jeremy Gow

Comparing Mathematical Provers 188
    Freek Wiedijk

Translating Mizar for First Order Theorem Provers 203
    Josef Urban

Invited Talk

The Mathematical Semantic Web 216
    Massimo Marchiori

Author Index 225

Digitisation, Representation, and Formalisation
Digital Libraries of Mathematics

Andrew A. Adams*

School of Systems Engineering, The University of Reading.
A.A.Adams@Rdg.ac.uk

Abstract. One of the main tasks of the mathematical knowledge management community must surely be to enhance access to mathematics on digital systems. In this paper we present a spectrum of approaches to solving the various problems inherent in this task, arguing that a variety of approaches is both necessary and useful. The main ideas presented are about the differences between digitised mathematics, digitally represented mathematics and formalised mathematics. Each has its part to play in managing mathematical information in a connected world. Digitised material is that which is embodied in a computer file, accessible and displayable locally or globally. Represented material is digital material in which there is some structure (usually syntactic in nature) which maps to the mathematics contained in the digitised information. Formalised material is that in which both the syntax and semantics of the represented material, is automatically accessible. Given the range of mathematical information to which access is desired, and the limited resources available for managing that information, we must ensure that these resources are applied to digitise, form representations of or formalise, existing and new mathematical information in such a way as to extract the most benefit from the least expenditure of resources. We also analyse some of the various social and legal issues which surround the practical tasks.

1 Introduction

In this paper we present an overview of the use of information technology in mathematics. Some of the earliest uses of computers was the automation of mathematics. Babbage's Difference and Analytical Engines [Bab23] were designed to engineer tedious, error-prone, mathematical calculations. Some of the first digital computers were designed for the purpose of breaking encryption codes. In the early days of digital computing, much was expected of this mechanical revolution in computation, and yet so little appeared for decades. As with the Strong AI and Ubiquitous Formal Methods communities, early over-selling of the idea of mechanised mathematical assistants led to a dismissal of the whole idea by many in the mathematics community.

* This work is supported by EU Grant MKMNet IST-2001-37057 and UK EPSRC Grant GR/S10919

A. Asperti, B. Buchberger, J.H. Davenport (Eds.): MKM 2003, LNCS 2594, pp. 1–16, 2003.

This raises a side question as to who is part of this "mathematical community". Athale and Athale presented a modest categorisation of this community in [AA01]. In the sense of developing tools and techniques for mathematical knowledge management(MKM), we must be as inclusive as possible in defining this community. The narrower our definition, the less use our work will be, and the more likely to become a sideline in the progress of information technology. That is not to say that the MKM community should not define specific tasks with narrow, achievable goals. Far from it, but for each of these goals, the needs and milieu of the whole mathematical community should be considered before fixing on the specifics of a project. Some projects will, naturally, only be of interest and benefit to a specific small group within the mathematics community. Even these, however, will benefit from a broad strategic overview being used to inform their development. The general and the specific must always be kept in balance when developing technology. Lack of this overarching viewpoint has led to the fragmentation of mathematical tools we see currently. This paper presents a wide-ranging view of who the users of MKM software might be, their needs and what approaches appear fruitful for further exploration.

We will begin by considering the various levels on which information may be represented on computers: Digitisation (for display) in Sect. 2; Representation (purely syntactic detail) in Sect. 3; and finally Formalisation (both syntax and semantics included) in Sect. 4. Once these categories of mathematical information have been described in detail, we will turn our attention to the uses of this information: in Sect. 5, to the dissemination of mathematical information to the various types of user; in Sect. 5.1, to the methods of dissemination historically and to the present; to some ideas on future directions in Sect. 5.2; lastly to legal and social issues effecting MKM in Sect. 5.3. In Sect. 6 we survey prior efforts in aspects of MKM, many of which form the background to the work of current researchers in the community, and which will form background or components for our developing technologies. We finish with some concluding remarks in Sect. 7.

2   Digitisation

New mathematical material is being produced by digital means. In many subjects this would lead to knowledge management specialists for that field discounting the necessity of spending much effort on bringing the older non-digital material into the digital world. In mathematics, however, the effort would not be misplaced. Even given the ever-increasing rate of publication of new mathematics (estimated to be rising exponentially at present: the total amount of mathematics that has been published appears to be doubling every ten years). This compares with the science in general where it is estimated that only the number of new works published every ten years doubles (as opposed to the total published) [Kea00]. Mathematics is a very hierarchical subject. It is rare that new work completely supersedes older work. Unlike physics, for example, where new theories expand upon and quite often replace older work, leaving the older to languish in the annals of the history of science, older mathematics is rarely completely replaced. Newer proofs may be shorter or more elegant, even appearing to approach some divine perfection [AZ99]. However, when a new area of study opens up it is often the case that older proofs, in their less refined and honed state, may give more insight into approaches, than the newer subject-specific ones. So, mathematics, and mathematical knowledge, is rarely, if ever, superseded. See [Mic01] for an interesting look at a specific digitisation effort.

Given this idea, that older mathematics may be as useful as the latest up-to-date material, it is imperative that efforts at MKM do not begin from 1962 (LISP 1.5 [McC62]); 1979 (TEX [Knu79]); 1988 (Mathematica [Wol])...and work forward with existing represented and formalised mathematics. (In Sect. 3, we will discuss the work on automatic translation of digitised mathematics texts into a representational format.) Seminal works may still go out of print, or be available from small publishers and almost unobtainable, try to buy a copy of [ML84] for instance. University libraries periodically cull their stock, and printed matter is always susceptible to accidental destruction. Material is still subject to copyright and, as we shall discuss in Sect. 5.3, the copyright holders may not even be the original authors. For instance, I am reliably informed that photocopies of the difficult-to-obtain (supposedly still in print) [ML84] are available. While this may be true for widely-acknowledged seminal works, it is less likely to be true for a wide range of material, languishing in the hands of uninterested publishers refusing to relinquish rights gained from highly biased contracts. A project such as Project Gutenberg [Har71] should be created as part of a large-scale investigation into MKM, not to transcribe works of mathematics, but simply to capture their graphic essence, to guard against loss and degradation of the physical originals.

That is not to say that digital materials are themselves completely proof against loss and degradation. [JB01] includes many details on the general issues, following research by and for The British Library on digital collections. Preservation of material in digital form requires a different approach to the preservation of physical material. Some issues are easier to solve for digital copies than physical ones: digital copies, when made carefully, are identical to the original and further copies of the copy also do not degrade. Other issues are more tricky in the digital world: language changes over time may make the "format" of a physical copy difficult to process, but this usually takes centuries to become a serious problem (except in the case of rapidly disappearing languages), whereas the formats of binary files may change very quickly and backwards compatibility between formats may not be preserved even for years, let alone decades or centuries. The physical form of the digital data must also be considered in terms of potential damage or outdated storage mechanisms. The University of Edinburgh, for instance, donated their last punched card reading machine to a computer museum in the 1990s. A year or so later the University had to ask for it to be loaned back because a number of members of the Physics and Chemistry department had data that would be expensive to reproduce stored on punched cards in filing cabinets. CD Roms have only a limited shelf life and although backwards compatibility with CD Roms is currently being maintained this may not be the case for decades longer. We will discuss other issues related to this in the section (3) on represented mathematics.

Current work on the digital display of mathematics focusses on how to efficiently transmit mathematical data which is already digitally represented, and how to coherently display that mathematics to capture the intent of the author while fitting with the restrictions of the reader's display method. While an issue for MKM, this is much less important (and recent work has been very successful at addressing these issues) than the reverse engineering one of taking displayed mathematics, commonly a mixture of consistent and inconsistent printing formats, and creating a digital archive, which may then be amenable to further processing, which we now consider.

3   Representation

The problem of cross-communication between differing systems of formalisation of mathematics has received a moderate amount of interest over the last decade or so. We will address this issue in detail in Sect. 4. In this section, we concentrate on the gap between formalised mathematics and digitally displayed mathematics, which is the simple representation of the syntax without necessarily linking this to an underlying semantics. Davenport's overview [Dav01] provides useful background to both this and the following section.

Typeset mathematics comprises two primary types of text: the mathematical text and the explanatory text. The richness of the combination is one of the causes of the massive expansion factor involved in fully formalising a published mathematics text. Modern mathematics may be typeset in a variety of systems. Some of these systems include some level of formalisation supporting the mathematical text (e.g. in a Mathematica Workbook). Others, such as LaTeX [Knu79, Lam94] or Display-MathML [Fro02], only contain typesetting information which may be partially converted to a formal semantic scheme. (We will consider representations which include fully or partially formalised semantics in Sect. 4.) There are two primary sources of represented mathematics: papers produced by mathematics users/producers for which the original encoding is available (as opposed to merely the graphical output); and scanned mathematics papers. We will consider these two types of information separately as there are some very different issues to consider. After this, we will return to the connection between the mathematical text and the explanatory text.

3.1   Representation from Digital Encoding

Some interesting work has been done on recovering some of the author's *mathematical intent* from the printed or digital graphic format of published mathematics. A very interesting paper is that of Fateman et al [FTBM96]. The main issues in this topic are the recognition of symbols in the rich language of mathematics, and the parsing of the two dimensional presentation used in typesetting. Take the simple instance of integration of real-valued functions. An integral presented in typeset form might look like this:

$$\int_a^b f(x)\mathrm{d}x$$

or this: $\int_p^q f(y)\mathrm{d}y$. These may have identical semantics, yet are displayed somewhat differently. Mechanical parsing of these two-dimensional graphical languages poses problems that have received limited attention from the optical recognition community (see comments in [FTBM96] for details. Since recovery of representational material from existing graphical representations (whether scanned or original digital documents) is important to the field of MKM, it is up to us in the MKM community to seek collaboration with those working in the area of parsing of graphic images to develop this field.

Without a suitable underlying representation language in which to embed the translations, however, the utility of parsing techniques for graphical images is entirely pointless for a large scale effort at enabling access to a wider corpus. We will return to this point in Sect. 4.1, where we will consider current systems which may be good candidates for such representation.

3.2   Representation in the Primary Source

Representations which are available with the source code (e.g. a LaTeX file) are already in a form suitable for automatic parsing to add in some level of formalisation. Even a veneer of formalisation, providing it is accurately and automatically produced, can be highly useful in the task of MKM. However, challenges still remain. Authors use their own style files and redefine commands. Some intelligence must be applied to the parsing of even a well-written LaTeX file, since the aim of the author is to present the text clearly in the graphical output to a human reader, rather than to produce machine-parseable code. This same issue attaches to other methods of typesetting mathematics to a greater or lesser degree: word processor files using an equation editor; papers written using a Computer Algebra System (CAS); etc. This issue is also related to that of the explanatory text, and its relationship to the mathematical text. While inline formulae are obviously connected to the sentence that surrounds them, English is full of hanging anaphora, even in the most clearly written technical paper. Cross-references to equations and lemmas buried in other parts of the paper, in other papers, and inside floating tables and figures, further cloud the issue. Even with the abstraction layer of \label and \ref in a LaTeX file, such internal links may be obscure and are almost certainly incomplete.

Some very good initial work on these kinds of problems can be found in [BB01], which looks at the problem of developing new documents from a variety of originals, through the use of text-slicing. Some automation is present, but most of the '"meta-information" (information about the document) is hand-crafted. Beyond this initial input of meta-information, however, there is an automated layer which ties a variety of documents together in a coherent form. The purely practical issues of converging a variety of sources has also been considered in this project.

3.3   Representation and Explanation

The relationship between the natural language parts of a mathematics text and the symbolic content is further complicated by a variety of nomenclatures and symbol sets. The older the mathematical text we wish to represent in a manipulable form, the wider the deviation is likely to be from current standard usages. Given that this is already a problem requiring the application of intelligence to understand the context of the pure mathematical text, it is unlikely that current automatic tools will be able to solve this problem. Automatic cross-referencing within a paper can help but we must accept the limitations of current technology and not expect at this stage too fine grained an automatic analysis of maths texts.

One area which should be relatively easy within this area, however, is to cross-link documents both forwards and backwards via citation indices. Each mathematical document in a digital mathematics library should have an associated set of links pointing to papers which reference it and which are also available in the library. Likewise links to the papers referenced from a work should be present, both where they appear in the text and from the bibliography section at the end. See Sect. 5.4 for further discussion of cross-referencing to external items.

3.4   Maintenance of Archives of Represented Material

As previously mentioned, the mere fact that data is digital is not a guarantee of its incorruptibility. Some basic precautions, such as adequate backup facilities, are obvious, and the process of mirroring data worldwide is another step int he right direction. However, the subject of maintaining digital archives is a current topic of much discussion amongst library scientists, and the MKM community should take note of such developments. Beyond simple data hygiene and archival storage of snapshots of data, there is a definite and growing problem in formalised mathematics that the amount of effort needed to keep formal developments up to date with the growing, and therefore changing, capabilities of the formal systems is increasing. If the required effort increases beyond the funding available for maintenance on such projects then there is a substantial risk of the prior work degrading, possibly degrading beyond the point where it is quicker and cheaper to recreate the lost data when it is needed than to try and bring it up to the state of the art.

There are some interesting developments in this area coming from the NuPrl group at Cornell University. Growing from a perceived need to maintain large formal developments of abstract and applied mathematics with their own systems, the NuPrl [CA+86] group developed a "librarian" architecture which maintains not only the current proof status of objects with reference to the "version" of the system, but also attempts to patch proofs that fail with new versions using automated deduction techniques. The system also maintains records and copies of the older versions of the system in which earlier developments are valid, and documents that changes between versions. This provides a framework for keeping developments valid with as up to date a version of the system as is feasible given limited resources. Following on from this development, the Office of Naval Research has funded a project at Cornell to produce a "library of digital mathematics" to support "the production of mathematically proved programs". This has consisted of integrating various other theorem proving and some other specialist mathematical systems within the same "librarian" framework. In addition, a fair amount of work has been done on reconciling the differences between the underlying logics of these various systems (HOL, NuPrl, MetaPRL, PVS, Coq [BB+96], Isabelle [Pau88]...) where possible, and on identifying the difference where not. Thus, a theorem proved in one system may be identified as compatible or incompatible with the basic logic of another (such as the incompatibility derived from one system being classical and another constructive in nature).

The development of mathematical software has reached the point of maturity where there are many users and constant small upgrades to the capabilities of the software. It has also reached the point where it is usable not only to experts int he field but to users from other fields wishing to benefit from the advances. This is a crucial juncture in the development of a field like this and efforts in preserving a legacy now should prevent much wailing and gnashing of teeth in the future over incompatible and out of date archives having to be reinvented instead of merely updated.

4   Formalisation

Formalised mathematical developments might be thought to be the easiest to control and make use of, given their highly structured nature, and the embedded semantics necessary for their development. This semantics, however, is purely an operational semantics. To be useful in supporting human mathematical endeavour, some form of denotational semantics or Kripke semantics is still needed to connect the type `Real` in PVS [SOR] with the type `Real` in HOL [GM93], and to distinguish it from the type `Real` in Mathematica [Wol] which is actually a floating point representation.

A note of caution about management of knowledge in formalised mathematics. As the body of work in the various systems (PVS, HOL, Mathematica etc.) progress, variations on a theme become more likely. As the body of work in these areas grows, the chances of re-inventing the wheel become ever more apparent. Large formalisations run to thousands of lemmas and definitions, and despite efforts at consistent naming, the sheer number of variants required for a formalisation leads to a lack of clarity. Textual searching through the formal development, while more easily achieved than textual search through less structured representations of mathematics, can still lead to an overload of possible results. So, we must not ignore the needs of the user communities of these systems for good search mechanisms based on higher order matching and some form of semantic matching.

4.1   Representation Languages and Formats

There are a number of possible frameworks for use in storing fully formal representations of mathematics. Some of these are briefly discussed in Sect. 6 on Prior Art which informs and provides a foundation for MKM. Here we will compare some of the existing systems in their scope and utility. A full exploration of these varied systems would require at least one, probably many, papers so by necessity this is only a brief overview.

OpenMath [Dew00b,Dew00a] and OMDoc [Koh01] form an interesting starting point for consideration of formats for representing mathematics. As pointed out in [APSCS01,APSC+01], however, the source of OpenMath as primarily a transport layer for CAS, later expanded to cover theorem provers, leaves some aspects of it unsuitable as a generic formal representation system for mathematical information.

There are a number of individual systems with a good underlying format for the representation of parts of mathematics, such as Mizar [Try80]. However, these systems tend to be tied very strongly to a particular system and a particular viewpoint of mathematics. SO, while these systems are worthwhile and form a highly useful and usable source of mathematical knowledge, it is unlikely that they will be useful as a format for generic representations.

The same can be said of the various theorem provers and computer algebra systems. Their languages, and even more cross-system developments such as the CASL [Mos97] are still more focussed on a single kind of mathematical knowledge.

In bringing MKM into sharper focus there must be a review of the various languages in terms of their capabilities, their current usage and their suitability for the needs of MKM. This must be done in parallel with identifying requirements and useful goals in the type of information needed to be accessed, and how that information can be described in a formal way as well as stored in a formal way.

5   Information Dissemination

So, we have seen that the various types of mathematics held on and accessible by computer have their problems. The job of MKM, like that of a librarian, is not just to catalogue, file and cross-reference information. It is to provide it to those who desire it, in as usable a form as possible. Hence in this section we consider the problem of MKM in terms of the people involved, their needs and the possible ways MKM might meet their needs.

5.1   Different Users and Their Needs

When considering types of user for MKM, we are not considering individuals and splitting them into discrete sets. Rather we are (to use a term from the Human-Computer Interaction (HCI) community) considering the "modes" of working displayed by users at different times. These modes can be categorised as:

Producer creating new mathematical information.
Cataloguer filing and sorting a variety of mathematical information into new collections.
Consumer using mathematical information in an application, or for educational purposes.

Note that in the process of producing mathematical information, most will also be consuming other pieces: it is a rare mathematician who does not depend on the body of mathematics built up over two thousand years, at least by direct reference to prior art. A cataloguer is a special form of producer, but one with such an obviously key role in MKM that we feel that mode should be singled out for more detailed attention (see Sect. 5.4 in particular). Finally, even in the "simple" act of applying prior art, a consumer may still produce new material worth being disseminated to others.

The other useful distinction between classes of user is in their type of employment. Mathematicians have always included highly skilled "amateurs" in their ranks as well as those professionally involved in the production of mathematics. These, and those primarily involved in pure academic positions have specific goals, generally involving as wide as possible a dissemination of their work.

Non-mathematician academics in a wide variety of fields use mathematics and produce their own brand of mathematical developments. Computer scientists, engineer, economists, archaeologists and many others contribute to and benefit from a wide range of sources of mathematical information, and need good MKM for the development of their field.

Industrial mathematicians may also be interested in wide distribution of their own work (for instance see the notes on Intel in 5.3 below), or may be interested in allowing paid access to parts of it. They also desire access to the work of others to improve their own developments.

These different types of user also have different modalities of operation. All, at one time or another, can be regarded in the mode of "learner" or "student". Of course, this includes the user for whom this is the primary (sometimes sole) mode, the student of mathematics. Following on from the student mode of learning a piece of mathematics, there is the "application" mode, where the user is getting to grips with a piece of mathematics and using it to solve a problem. Lastly there is the "development mode" where the user understands the mathematics well and may come up with novel twists or uses for it, or create entirely new concepts based on the original.

5.2   From Caxton to CERN

From the development of movable type through to recent developments in MathML and on-line journals, the dissemination of mathematics has gone through many changes. Originally one could access such material only by talking to the mathematician who originated a new concept, or via a limited number of libraries containing scrolls and then books laboriously copied by hand, an error-prone procedure, especially when carried out by those with no mathematical training themselves. The days of movable type improved this process, but the expense of professional typesetting of mathematics, together with personal typewriters and early word-processors (with type limited to a few symbols beyond alphanumerics) led to a large number of monographs, theses and the like being produced with typed words but hand-drawn symbology. Gradually this gave way to more powerful digital solutions and word processors with equation editors, TEX and LATEX, and CAS with various output options.

The business of publishing has gone through many similar changes. In the beginning natural sciences were supported by the church and universities (many of them supported by the church as well, though some more state-oriented institutions existed quite early on). Movable type in particular led to the development of a business model whereby a publisher put up the costs of the printing (either directly by ownership or by paying a separate printing business) and took a share of the profits gained from distributing the product. Books and journals on mathematics have proliferated over the years, and the advent of cheap self-publication on the Web has led to changes in both the business models and the academic models. However, these models are slow to change and technology has overtaken the speed with which slow evolution can happen. Pressures are building in both the academic and business worlds which MKM must take account of in order to help rather than hinder the goal of MKM, which is to better disseminate mathematical information to those who need it, while allowing due benefit to those who produced it.

5.3   Copyright, Corporate Publishing, and Quality Concerns

In recent years the academic publishers, once concerned with the quality of their output at least as much with its profitability, have mutated out of recognition. The "independent" publishers have mostly been bought up by large multi-media conglomerates and despite the efforts of those who care about the content, the model of the academic journal, edited and refereed by unpaid academics for the prestige they gained, is becoming less and less viable with respect to these large conglomerates, who have priced themselves out of the market in many cases, by increasing the costs to libraries well beyond the increase in University funding in most places. Increasingly, academics are beginning to wonder what these costs are doing other than expanding the bottom line of the publishing houses. Even the University publishing houses are under pressure from their parent institutions to produce more income, as other sources decrease. (This is not true everywhere, but even where the institutions are not under financial pressure, the profits made by the non-University houses leads to similar increases at the University houses in many cases.)

The unpaid labour provided by academics and scholarly societies is increasingly begrudged. The ever more restrictive copyright terms demanded by publishers is another cause for concern, leading the International Mathematics Union to recommend a counter-proposal for copyright agreements [Bor01]. One leading mathematics academic has even announced that he had abandoned reading journals in favour of the Alamos archive of "live" papers. As academics face pressure to produce more, the time and energy they have for the unpaid work of quality control on the publications of their peers continues to diminish. MKM surely has a role to play in improving the efficiency of this work, allowing faster, more insightful refereeing within new models of publication. This needs to be true not just for formal mathematics, which is supposed to automate some of the refereeing process by using a proof checker to test the validity of claims, allowing the referee to study the definitions and final resulting theorems rather than needing to work through the intermediate result. The role of a referee can be seen as part of the cataloguing work that needs to be done for the efficient management of mathematical information. The role of referees in determining the importance and overall quality of mathematics that is a candidate for publication would be very difficult to automate, but the book-keeping involved in this process can be made much easier with the correct attribution of meta-data, and the easy availability of referenced prior art. The recent scandal of falsified results in physics at Bell Labs [Bel02], with papers published in Science and Nature, reinforces the necessity of providing good support for the peer review system in mathematics and related subjects.

The attitude of the free software movement, that better code comes from free distribution and the right to produce derivative works, has many similarities with the attitudes of mathematicians (particularly academics). Even companies who business model is built primarily on the value of certain classes of intellectual property (such as Intel Corporation) recognise that other intellectual developments (such as the formalised mathematics of Harrison [Har00]) are more valuable in free dissemination than in closed-source exploitation. Formalised mathematics still mostly follows this ideal and MKM tool developers need to be aware of the desires of the mathematics communities in similar respects.

The idea of the mathematical toolbox, like the IT toolbox being developed as Open Source Software, can be shown to benefit a variety of users. Mathematics, like IT, is more often a tool in wealth generation, than the product generating the wealth directly. it is in the interests of society and mathematicians that the tools in these toolkits are as available as possible, and flexible in their applicability.

So far in this section we have focussed on the issue of new mathematics being published. As we have said before, however, MKM is at least as much about enabling access to the existing corpus as it is about new material. Here we have a difficult problem, posed by existing agreements between publishers and the original authors. While there are many possible copyright agreements between author and publisher, the ones which concern us are those which either assign copyright (or all rights of exploitation) to the publisher, or those which restrict electronic rights to the author themselves. For material which needs digitisation (from journals, books or conference proceedings) some of this is probably now out of copyright (though less and less new material is being added to the intellectual commons in this way due to ever-increasing lengths of copyright protection). However, the vast majority of material published after 1930 will be under copyright still, and care would have to be taken with any large-scale digitisation effort to comply with the appropriate law (especially with international treaties since such an effort would almost certainly be undertaken in various global locations). Any estimates of the effort involved in such work should include costings on contacting, negotiating with, and possibly paying, copyright holders for the right to transfer their material to a new form (i.e. producing a derivative work). It should be noted that even where an author has not signed over copyright in the work itself to the publisher, the publisher still retains copyright over the layout of the material and must be included in discussions.

5.4   Meta-information

The Web in general is beginning to suffer from information overload. Legal arguments about deep-linking, the copyright status of "collections of information" and digital copyright laws are contributing to the chaos in general information management at present. While it can be expected that these arguments will go on for some time, it can be hoped that relatively stable situations can be reached with respect to parts of the mathematics community quite quickly.

In respect of this, it is in the development of good infrastructure for supporting meta-level information (information *about* a piece of digital mathematics) that MKM may have the quickest and most prevalent effect. Meta-Information can be produced automatically or "manually", and in the case of information only digitised but not represented, manual production is quite probably the only feasible way forward. Some of this meta-information already exists, in fact, such as the straight citations of papers. As we shall see below, however, even these are not without problems are present. As with much of MKM development, the limited resources available must be put to good use, and frequently, the manual or machine-assisted production of good quality meta-information will be of far more use than the translation of digitised material into symbolic representations. Only once automatic translation and automatic extraction of meta-information become possible will this change. All these areas should be explored and developed but the near-term realisable goals should be pursued with greater vigour in the short term while laying the groundwork for improved automation in the longer term.

While MathSciNet [AMS], the Alamos archive, citeseer [Cit] and other systems are useful, the lack of a good underlying representation of the mathematical content of pages reduces that utility exponentially in the number of papers stored. Note, for instance, that most journals in mathematics and the mathematical sciences, such as computing and physics, generally advise authors to avoid use of mathematical formulae in the titles of papers, so that text-only searching can be applied at that level. Good symbolic representations and formalisations are obviously the way in which meta-information should be stored an manipulated, and it is in this area that the experience of those working in theorem proving (automated and machine-assisted) and CAS should prove highly useful. HCI experts are also important to such undertakings. An excellent database of meta-information about digitally stored mathematics remains useless if the interface is counter-intuitive. Not only must the underlying representation of meta-information be suitable for automatic processing and searching (a hard problem in itself) it must also be possible to enter and display the queries and results in a usable fashion. The experience of HCI experts in theorem proving and CAS should again be invaluable in this arena.

The various technologies mentioned above (OpenMath, OMDoc, HELM, "book-slicing") are a vital starting point for identifying suitable technologies for storing and accessing mathematically useful meta-information. Once good methods for storing and entering meta-information for mathematics have been developed, there needs to be a concerted effort to "sell" it to as many producers and cataloguers of mathematics as possible. This does not mean ask them to part with money for the right to be involved in the web of knowledge, but to persuade them that being involved is in their interests and that making their meta-information freely available to others in return for reciprocal access, will benefit them in the long run. The intellectual property issues, in both a legal and a social sense, make this area a sensitive one that must be handled with care.

6   Prior Art: Previous and Current Projects of Relevance to MKM

It would be impossible to cover all the projects that have led to the formation of the current MKM community. However, there are various projects, either recently completed or currently ongoing, to which members of this community should pay close attention, particularly to the projects they themselves were not involved with, the assumption being that they are well aware of the potential contributions of their own previous work. So, in this section we will give a quick description of a number of project, primarily EU funded, which form the background to current developments of a concerted effort in Europe, not ignoring the contributions of those elsewhere, particularly in the US. Some of these projects have already been discussed in more detail but it is useful to have them mentioned here again.

- **Monet: http://monet.nag.co.uk/**
  Mathematics on the Net. A recently funded EU project starting in early 2002. This project is aimed at the general theme of "mathematical web services", generic mathematical services offered over the internet.
- **OpenMath: http://monet.nag.co.uk/openmath**

  A Common Mathematical representation language for mathematical software, particularly CAS and theorem proving systems. The OpenMath Society continues to support and extend the standard. A new EU project started in 2001, following on from the original project which ran from 1997 to 2000.
- **Calculemus: http://www.calculemus.net/**
  A project to combine the capabilities of CAS (excellent for calculation) and Theorem Provers (good for proving logical statements) into better mathematics software. In addition to an EU funded network of sites with junior researchers, there are annual workshops/conferences and a recent autumn school.
- **MoWGLI: http://www.mowgli.cs.unibo.it/**
  Mathematics on the Web, Get it by Logic and Interfaces
  The MoWGLI abstract from their website:
  The World Wide Web is already the largest resource of mathematical knowledge, and its importance will be exponentiated by emerging display technologies like MathML. However, almost all mathematical documents available on the Web are marked up only for presentation, severely crippling the

potentialities for automation, interoperability, sophisticated searching mechanisms, intelligent applications, transformation and processing. The goal of the project is to overcome these limitations, passing from a machine-readable to a machine-understandable representation of the information, and developing the technological infrastructure for its exploitation. MoWGLI builds on previous standards for the management and publishing of mathematical documents (MathML, OpenMath, OMDoc), integrating them with different XML technologies (XSLT, RDF, etc).

- **Euler**

  Euler is a relatively recently funded EU project to develop a web portal for mathematics available on the web. This project is aimed at the user interface level of searching for, and using, mathematical information and services via the internet.

- **Mizar**
  One of the oldest Formalised Mathematics projects. Since the early 1970s the Mizar project has been a central point for a large development of fully formalised mathematics and the development of the Mizar system and its components. Various aspects of the system have influenced the development of related projects, such as the ability to produce "human-readable" versions of proofs.

- **MKMNet: http://monet.nag.co.uk/mkm/**

  This EU 5th Framework project (IST-2001-37057) has been funded to support the development of a 6th Framework Integrated Programme for Mathematical Knowledge Management. It is run by a consortium of (primarily) academics most of whom were present at the MKM '01 workshop.

- **NIST Digital Library of Mathematical Functions**

  Based on the *Handbook of Mathematical Functions* [AS72], this NIST project aims to digitise and extend the utility of this seminal mathematical reference text. Two papers were presented about this project at MKM '01: [Loz01, MY01].

- **Digital Library of Mathematics: Cornell University**

  This project has already been described in some detail in Sect. 3.4. it aims to produce a library of formalised mathematics, from various theorem proving and related systems, for use in program verification and synthesis, for the US ONR.

7   Conclusions

We hope we have covered a wide range of the issues surrounding MKM in an interesting and enlightening way. The differing viewpoints of technical issues (the differentiation and the links between digitised, symbolically represented and formalised mathematics) and user issues (copyright, HCI, mate-information) have been covered and some of the links between them explored. This is something of a position paper, and some of the ideas are drawn from the discussions since the first MKM workshop in Linz [BC01].

Attention must be paid not only to the technical issues but the social and legal ones as well. Without the engagement of as many of the stakeholders as possible, the development of MKM will be held back.

The role of mathematics in the world continues to grow. While computers seem ubiquitous, it is really mathematics that is ubiquitous, from e-science initiatives, to cryptographic protocols, they're all based on mathematics as well as computer hardware. Without good MKM, the development of knowledge economies and a net-enabled world is delayed and made poorer.

Acknowledgements. Many of the ideas in this paper are refined from discussions with members of the MKMNet consortium and attendees at MKM '01. I would also like to thank the anonymous referees for a number of good suggestions including the list of related EU project which have contributed to the development of the MKMNet project.

References

[AA01]   M. Athale and R. Athale. Exchange of mathematical information on the web: Present and Future. In Buchberger and Caprotti [BC01].

[AMS]   AMS. Mathscinet. www.ams.org/mathscinet/.

[APSC⁺01]   A. Asperti, L. Padovani, C. Sacerdoti Coen, G. Ferruccio, , and I. Schena. Mathematical Knowledge Management in HELM. In Buchberger and Caprotti [BC01].

[APSCS01]   A. Asperti, L. Padovani, C. Sacerdoti Coen, and I. Schena. HELM and the Semantic Math-Web. Springer-Verlag LNCS 2152, 2001.

[AS72]   M. Abramowitz and I. A. Stegun. *Handbook of mathematical functions with formulas, graphs and mathematical tables.* Dover, 1972.

[AZ99]   M. Aigner and G. M. Ziegler. *Proofs from THE BOOK.* Springer, 1999.

[Bab23]   C. Babbage. On the Theoretical Principles of the Machinery for Calculating Tables. *Edin Phtl Jrl,* 8:122–128, 1823.

[BB⁺96]   B. Barras, S. Boutin, et al. The *Coq* Proof Assistant Reference Manual (Version 6.1). Technical report, INRIA, 1996. Available on-line with *Coq* distribution from ftp.inria.fr.

[BB01]   P. Baumgartner and A. Blohm. Automated Deduction Techniques for the Management of Personal Documents (Extended Abstract). In Buchberger and Caprotti [BC01].

[BC01]   B. Buchberger and O. Caprotti, editors. *MKM 2001 (First International Workshop on Mathematical Knowledge Management).* http://www.risc.uni-linz.ac.at/conferences/MKM2001/Proceedings, 2001.

[Bel02]   Research Review, 2002. http://www.lucent.com/news_events/researchreview.html.

[Bor01]   J. M. Borwein. The International Math Union's Electronic Initiatives (Extended Abstract). In Buchberger and Caprotti [BC01].

[CA⁺86]   R. L. Constable, S. F. Allen, et al. *Implementing Mathematics with the NuPrl Proof Development System.* Prentice-Hall, 1986.

[Cit]   Citeseer. www.citeseer.org.

[Dav01]   J. Davenport. Mathematical Knowledge Representation (Extended Abstract). In Buchberger and Caprotti [BC01].

[Dew00a]   M. Dewar. OpenMath: An Overview. *ACM SIGSAM Bulletin*, 34(2):2–5, June 2000.

[Dew00b]   M. Dewar. Special Issue on OPENMATH. *ACM SIGSAM Bulletin*, 34(2), June 2000.

[Fro02]   M. Froumentin. Mathematics on the Web with MathML. http://www.w3.org/People/maxf/papers/iamc.ps, 2002.

[FTBM96]   R. Fateman, T. Tokuyasu, B. P. Berman, and N. Mitchell. Optical Character Recognition and Parsing of Typeset Mathematics. *Journal of Visual Communication and Image Representation*, 7(1):2–15, March 1996.

[GM93]   M. J. C. Gordon and T. F. Melham, editors. *Introduction to HOL*. CUP, 1993.

[Har71]   M. Hart. Project gutenberg, 1971. www.gutenberg.org.

[Har00]   J. Harrison. Formal verification of floating point trigonometric functions. Springer-Verlag LNCS 1954, 2000.

[JB01]   M. Jones and N. Beagrie. *Preservation Management of Digital Materials (A Handbook)*. The British Library, 2001.

[Kea00]   T. Kealey. More is less. *Nature*, 405(279), May 2000.

[Knu79]   D. Knuth. *TEX and METAFONT: New directions in Typesetting*. AMS and Digital Press, 1979.

[Koh01]   M. Kohlhase. OMDoc: Towards an Internet Standard for the Administration, Distribution and Teaching of Mathematical Knowledge. In J. A. Campbell and E. Roanes-Lozano, editors, *Proceedings of Artificial Intelligence and Symbolic Computation 2000*, pages 32–52. Springer LNCS 1930, 2001.

[Lam94]   L. Lamport. *LATEX: A Document Preparation System, 2/E*. Addison Wesley, second edition, 1994.

[Loz01]   D. Lozier. The NIST Digital Library of Mathematical Functions Project. In Buchberger and Caprotti [BC01].

[McC62]   J. *et al.* McCarthy. *LISP 1.5 Programmer's Manual*. MIT Press, 1962.

[Mic01]   G. O. Michler. How to Build a Prototype for a Distributed Mathematics Archive Library. In Buchberger and Caprotti [BC01].

[ML84]   P. Martin-Löf. *Intuitionistic Type Theory*. Bibliopolis, 1984.

[Mos97]   P. D. Mosses. CoFI: The Common Framework Initiative for Algebraic Specification and Development. pages 115–137. Springer LNCS 1214, 1997.

[MY01]   B. R. Miller and A. Youssef. Technical Aspects of the Digital Library of Mathematical Functions *Dreams and Realities*. In Buchberger and Caprotti [BC01].

[Pau88]   L. C. Paulson. The Foundation of a Generic Theorem Prover. *J. Automated Reasoning*, 5:363–396, 1988.

[SOR]   N. Shankar, S. Owre, and J. M. Rushby. *The PVS Proof Checker: A Reference Manual*. Computer Science Lab, SRI International.

[Try80]   A. Trybulec. *The Mizar Logic Information Language*, volume 1 of *Studies in Logic, Grammar and Rhetoric*. Bialystok, 1980.

[Wol]   www.wolfram.com.

MKM from Book to Computer: A Case Study

James H. Davenport*

Department of Computer Science, University of Bath, Bath BA2 7AY, England
J.H.Davenport@bath.ac.uk

Abstract. [2] is one of the great mathematical knowledge repositories. Nevertheless, it was written for a different era, and for human readership. In this paper, we describe the sorts of knowledge in one chapter (elementary transcendental functions) and the difficulties in making this sort of knowledge formal. This makes us ask questions about the nature of a Mathematical Knowledge Repository, and whether a database is enough, or whether more "intelligence" is required.

1 Introduction

It is a widely-held belief, though probably more among computer scientists and philosophers than among mathematicians themselves, that mathematics is a completely formal subject with its own, totally precise, language. Mathematicians know that what they write is in a "mathematical vernacular" [7], which could, *in principle* be rendered utterly formal, though very few mathematicians do so, or even see the point of doing so. In practice the mathematical vernacular is intended for communicating between human beings, or more precisely, mathematicians, or, more precisely still, mathematicians in that subject. The reader is meant to apply that nebulous quality of "common sense" when reading the mathematical vernacular.

It turns out to be remarkably hard to write "correct" mathematics in the mathematical vernacular. The problem is often with "obvious" special cases that are not stated explicitly[1], but which the knowledgeable reader will (and must) infer. There are even errors in [2] of this sort — see [11] and equation (23). The ninth printing of [2] contained corrections on 132 pages, and the Dover reprint of that corrected a further nine pages.

In this paper, we will explore the problems of representing a small part of one of the most well-known sources of mathematical knowledge: [2]. In particular, we consider the problems of translating the relevant content of chapter 4 — *Elementary Transcendental Functions* — into OpenMath [1,13]. In this paper we will be concerned with the semantic problems, rather than with the OpenMath

---

* The author was partially supported by the European OpenMath Thematic Network and the Mathematical Knowledge Management Network.

[1] The author recently had a problem with this, having set an examination question "if α is algebraic and β is transcendental, is $\alpha\beta$ always transcendental"? One students answered in the negative, quoting the case of $\alpha = 0$.

A. Asperti, B. Buchberger, J.H. Davenport (Eds.): MKM 2003, LNCS 2594, pp. 17–29, 2003.

problems. It should be noted that there is concern within the computer algebra community about the treatment of these functions in computer algebra [3,17].

It should be emphasised that this paper is in no way a criticism of [2]. The author is one of, probably literally, millions of people who have benefited from this enormous compendium of knowledge. Rather, the point is to illustrate that a book produced for other human beings to read, in days before the advent of (general-purpose) computer algebra systems or theorem provers, implicitly assumes knowledge in the reader that it is notoriously difficult to imbue such systems with. We therefore ask how we can make such knowledge explicit.

2   The "Elementary Transcendental Functions"

These days[2] these functions are normally considered to be exp and its inverse ln, the six trigonometric functions and their inverses, and the six hyperbolic functions and their inverses. For the purposes of this paper, we will class exp, the six trigonometric functions and the six hyperbolic functions together as the *forward functions*, and the remainder as the *inverse functions*.

The forward functions present comparatively little difficulty. They are continuous, arbitrarily-differentiable, many–to–one functions defined from $\mathbf{C}$ (possibly less a countable number of singularities) to $\mathbf{C}$. While it is possible to extend them to run from from the whole of $\mathbf{C}$ to $\mathbf{C} \cup \{\infty\}$, [2] sensibly chooses not to. The concept of ∞ is a difficult one to formalise (but see [4]), and, while $\mathbf{R} \subset \mathbf{C}$, it is not the case for their natural completions: $\mathbf{R} \cup \{-\infty, +\infty\} \not\subset \mathbf{C} \cup \{\infty\}$.

The problem lies rather with the inverse functions. They are continuous, arbitrarily-differentiable, one–to–many functions defined from $\mathbf{C}$ (possibly less a countable number of singularities) to an appropriate Riemann surface. The problem comes when we wish to consider them as functions from $\mathbf{C}$ (possibly less a countable number of singularities) to $\mathbf{C}$. The solution is to introduce "branch cuts", i.e. curves (though in practice we will only be considering lines in this paper) in $\mathbf{C}$ across which the inverse function is not continuous.

Provided that they satisfy appropriate mathematical conditions, any line or curve can be chosen as the branch cut. For example, ln, as one makes a complete counter-clockwise circle round the origin, increases in value by $2\pi i$. Therefore any simple curve from the origin to infinity will serve as a branch cut. The normal choice today[3], as in [2], is to choose the negative real axis.

It is also important to specify what the value of the function is on the branch cut. It clearly makes sense to have it continuous with one side or the other, and the common choice, as in [2], is to choose the value of ln on the branch cut to be continuous with the upper half-plane, so that $-\pi < \Im \ln z \leq \pi$. However, this choice is essentially arbitrary, and [16] would like to make the function two-

---

[2] Other trigonometric variants such as versine have disappeared. However, see section 5.

[3] Though the author was taught at school to use the positive real axis, with $0 \leq \Im \ln z < 2\pi$.

valued on the branch cut: $\ln(-1) = \pm\pi i$. This has the drawback of not fitting readily with numerical evaluation.

One still might wish to "have one's cake and eat it". [15] points out that the concept of a "signed zero"[4] [14] (for clarity, we write the positive zero as 0^+ and the negative one as 0^-) can be used to solve this dilemma, if we say that, for $x < 0$, $\ln(x + 0^+ i) = \ln|x| + \pi i$ whereas $\ln(x + 0^- i) = \ln|x| - \pi i$. However, this is no use to computer algebra systems, and little use to theorem provers.

The serious problem with branch cuts is that they make many "obvious" relations false. For example, exp takes complex conjugates to complex conjugates, as $\exp \overline{z} = \overline{\exp z}$, so one might expect the same, i.e.

$$\log \overline{z} \overset{?}{=} \overline{\log z}, \tag{1}$$

to be true of its inverse. Unfortunately, this is true everywhere except on the branch cut, where $z = \overline{z}$, and therefore $\log \overline{z} = \log z$. These complications mean that it is not a simple matter to codify knowledge about the inverse functions.

2.1 Encoding Branch Cut Information

[10] points out that most 'equalities' do not hold for the complex logarithm, e.g. $\ln(z^2) \neq 2 \ln z$ (try $z = -1$), and its generalisation

$$\ln(z_1 z_2) \neq \ln z_1 + \ln z_2. \tag{2}$$

The most fundamental of all non-equalities is $z = \ln \exp z$, with an obvious violation at $z = 2\pi i$. They therefore propose to introduce the *unwinding number* $\mathcal{K}$, defined[5] by

$$\mathcal{K}(z) = \frac{z - \ln \exp z}{2\pi i} = \left\lceil \frac{\Im z - \pi}{2\pi} \right\rceil \in \mathbf{Z} \tag{3}$$

We can then rescue equation (1) as

$$\ln \overline{z} = \overline{\ln z} - 2\pi i \mathcal{K}(\overline{\ln z}). \tag{4}$$

Since we know that $-\pi < \Im \ln z \leq \pi$, $-\pi \leq \Im \overline{\ln z} < \pi$. So the only places where the $\mathcal{K}$ term is non-zero is when $\Im \overline{\ln z} = -\pi$, i.e. $\Im \ln z = \pi$. Hence this equation implicitly encodes the region of invalidity of equation (1).

---

[4] One could ask why zero should be special and have two values (or four in the Cartesian complex plane). The answer is that all the branch cuts for the basic elementary functions (this is not true for, e.g. $\ln(i + \ln z)$, whose branch cut is $z \in \{e^t(\cos 1 + i \sin 1) \mid t \in (\infty, 0]\}$ are on either the real or imaginary axes, so the side to which the branch cut adheres depends on the sign of the imaginary or real part, including the sign of zero. With sufficient care, this technique can be used for other branch cuts as long as they are parallel with the axes, e.g. $\ln(z + i)$.

[5] Note that the sign convention here is the opposite to that of [10], which defined $\mathcal{K}(z)$ as $\lfloor \frac{\pi - \Im z}{2\pi} \rfloor$: the authors of [10] recanted later to keep the number of -1s occurring in formulae to a minimum.

3   Codifying ln

[2, p. 67] gives the branch cut $(-\infty, 0]$, and the rule [2, (4.1.2)] that

$$- \pi < \Im \ln z \leq \pi. \tag{5}$$

OpenMath has chosen to adopt equation (5) as the definition of the branch cut, rather than words, since it also conveys the necessary information about the value on the branch cut, which the form of words does not. From equation (5), one can deduce that the branch cut is $\{z \mid \Im \ln z = \pi\}$, which should be the same as $\{z \mid \Im \ln z = -\pi\}$. However, it takes a certain subtlety to convert this to $z \in (-\infty, 0]$, and maybe the branch cut should be stated explicitly, either instead of equation (5) (but then how does one specify the value on the branch cut?) or as well as it (in which case, how does one ensure coherence between the two?). However, despite the discussion in the previous section, precisely what formal semantics can one give to the phrase "branch cut"? Does it depend on one's semantic model for $\mathbf{C}$ and functions $\mathbf{C} \to \mathbf{C}$?

Currently, OpenMath does not encode equations such as equation (1) (since they are false). There are various options.

1. Encode them with unwinding numbers, as in equation (4).
2. Encode them as conditional equations, e.g.

$$z \notin \text{branch cut} \Rightarrow \log \overline{z} = \overline{\log z}, \tag{6}$$

3. Encode them via multivalued functions (see section 6)

The unwinding number approach is attractive, and it could be used in the "unwinding number approach" to simplification [8]. However, it would be useless to a system that did not support the semantics of unwinding numbers, though an "intelligent" database might be able to convert such an encoding into the conditional one. The conditional equation approach might be helpful to theorem provers, but the proof obligations that would build up might be unmanageable. In this form, it does not say what happens when z is on the branch cut, but an "else clause" could be added.

To state them in the "unwinding number" formalism, the following equations seem to be a suitable "knowledge base" for ln, in addition to equation (4).

$$\ln(z_1 z_2) = \ln z_1 + \ln z_2 - 2\pi i \mathcal{K}(\ln z_1 + \ln z_2). \tag{7}$$

$$\ln(z_1/z_2) = \ln z_1 - \ln z_2 - 2\pi i \mathcal{K}(\ln z_1 - \ln z_2). \tag{8}$$

The following is a re-writing of equation (3):

$$\ln \exp z = z - 2\pi i \mathcal{K}(z), \tag{9}$$

and we always have

$$\exp \ln z = z. \tag{10}$$

It is harder to write equations (7) and (8) in a "conditional" formalism, since what matters is not so much being on the branch cut as having crossed the branch cut. A direct formalism would be

$$(-\pi < \Im(\ln z_1 + \ln z_2)) \wedge (\Im(\ln z_1 + \ln z_2) \leq \pi) \Rightarrow \ln(z_1 z_2) = \ln z_1 + \ln z_2,$$

but, unlike equation (6), there is an input space of measure 0.5 on which this does not define the answer. One is really forced to go to something like

$$-\pi < \Im(\ln z_1 + \ln z_2) \leq \pi \Rightarrow \ln(z_1 z_2) = \ln z_1 + \ln z_2$$
$$\Im(\ln z_1 + \ln z_2) > \pi \Rightarrow \ln(z_1 z_2) = \ln z_1 + \ln z_2 - 2\pi i$$
$$\Im(\ln z_1 + \ln z_2) \leq -\pi \Rightarrow \ln(z_1 z_2) = \ln z_1 + \ln z_2 + 2\pi i$$

which is essentially equation (7) unwrapped.

3.1   Square Roots

It is possible to define $\sqrt{z} = \exp\left(\frac{1}{2}\ln x\right)$. This means that $\sqrt{}$ inherits the branch cut of ln. Since this definition is possible, and causes no significant problems, Occam's Razor tells us to use it. Equation (7) then implies

$$\sqrt{z_1 z_2} = \sqrt{z_1}\sqrt{z_2}\,(-1)^{\mathcal{K}(\ln z_1 + \ln z_2)}, \tag{11}$$

and the same discussion about alternative forms of equation (7) applies here. It is also possible to use the complex sign[6] function to reduce this to

$$\sqrt{z_1 z_2} = \operatorname{csgn}\left(\sqrt{z_1}\sqrt{z_2}\right)\sqrt{z_1}\sqrt{z_2}. \tag{12}$$

4   Other Inverse Functions

All the other forward functions can be defined in terms of exp. Hence one might wish to define all the other inverse functions in terms of ln. This is in fact principle 2 of [9] (and very close to the "Principal Expression" rule of [15]).

All these functions should be mathematically[7] defined in terms of ln, thus inheriting their branch cuts from the chosen branch cut for ln (equation 5).

---

[6] The csgn function was first defined in Maple. There is some uncertainty about csgn(0): is it 0 or 1, but for the reasons given in [6], we choose csgn(0) = 1.

[7] This does not imply that it is always right to compute them this way. There may be reasons of efficiency, numerical stability or plain economy (it is wasteful to compute a real arcsin in terms of complex logarithms and square roots) why a numerical, or even symbolic, implementation should be different, but the *semantics* should be those of this definition in terms of logarithms, possibly augmented by exceptional values when the logarithm formula is ill-defined.

In fact, it is not just the branch cut itself, but also the definition of the function on the branch cut, that follows from this principle, since we know the definition of ln on the branch cut.

[2] does not quite adhere to this principle. It does give definitions in terms of ln, but these are secondary to the main definitions, and, as in the case of [2, 4.4.26]

$$\text{Arcsin } x = -i \,\text{Ln}\left(\sqrt{1 - x^2} + ix\right) \qquad |x^2| \leq 1, \tag{13}$$

the range of applicability is limited. [15] suggested, and [9] followed, that equation (13) be adopted as the definition throughout $\mathbf{C}$. This has the consequence that

$$\arcsin(-z) = -\arcsin(z) \tag{14}$$

is valid throughout $\mathbf{C}$. No choice of values on the branch cut (compatible with $\sin\arcsin z = z$) can make $\overline{\arcsin(z)} = \arcsin(\overline{z})$ valid on the branch cut: it has to be rescued as

$$\begin{aligned}\overline{\arcsin z} = (-1)^{\mathcal{K}(-\ln(1 - z^2))} \arcsin \overline{z} \\ + \pi\mathcal{K}(-\ln(1 + z)) - \pi\mathcal{K}(-\ln(1 - z)).\end{aligned} \tag{15}$$

Here we have a fairly complicated formula, and the conditional form

$$(z \notin \mathbf{R}) \vee (z^2 \leq 1) \Rightarrow \overline{\arcsin z} = \arcsin \overline{z} \tag{16}$$

(which does not tell what happens on the branch cuts, but there $\overline{z} = z$) might be simpler.

For real variables, the addition rule for arctan can be written out conditionally [6]:

$$\arctan(z_1) + \arctan(z_2) = \arctan\left(\frac{z_1 + z_2}{1 - z_1 z_2}\right) \\ + \begin{cases} \pi & z_1 > 0, z_1 z_2 > 1 \\ 0 & z_1 \geq 0, z_1 z_2 \leq 1 \\ -\pi & z_1 < 0, z_1 z_2 \geq 1 \end{cases} \tag{17}$$

For both real and complex variables, there is a representation [8] in terms of unwinding numbers:

$$\arctan(z_1) + \arctan(z_2) = \arctan\left(\frac{z_1 + z_2}{1 - z_1 z_2}\right) + \\ \pi\mathcal{K}\left(2i(\arctan(z_1) + \arctan(z_2))\right). \tag{18}$$

It is also possible to write the law for addition of real arcsin of real arguments in a conditional form:

$$\sqrt{(1 - z_1^2)(1 - z_2^2)} - z_1 z_2 \geq 0 \Rightarrow \arcsin(z_1) + \arcsin(z_2) = A \tag{19}$$

$$\left(\sqrt{(1 - z_1^2)(1 - z_2^2)} - z_1 z_2 < 0\right) \wedge (z_1 > 0) \Rightarrow \arcsin(z_1) + \arcsin(z_2) = \pi - A$$

$$\left(\sqrt{(1 - z_1^2)(1 - z_2^2)} - z_1 z_2 < 0\right) \wedge (z_1 < 0) \Rightarrow \arcsin(z_1) + \arcsin(z_2) = -\pi - A,$$

where $A = \arcsin\left(z_1\sqrt{1 - z_2^2} + z_2\sqrt{1 - z_1^2}\right)$, but we have yet to find[8] an unwinding number formalism in terms of arcsin — there clearly is one in terms of (complex) lns, which works out to be $\arcsin(z_1) + \arcsin(z_2) =$

$$-i\left[\ln\left(iz_1\sqrt{1 - z_2^2} + iz_2\sqrt{1 - z_1^2} + (-1)^{\mathcal{K}(c_2)}\sqrt{1 - \left(z_1\sqrt{1 - z_2^2} + z_2\sqrt{1 - z_1^2}\right)^2}\right) + 2\pi i\mathcal{K}(c_1)\right],$$

where the correction terms are $c_1 = i(\arcsin(z_1) + \arcsin(z_2))$ and

$$c_2 = 2\ln\left(\sqrt{1 - z_1^2}\sqrt{1 - z_2^2} - z_1 z_2\right).$$

When $\mathcal{K}(c_2) = 0$, the main ln is recognisably $\arcsin\left(z_1\sqrt{1 - z_2^2} + z_2\sqrt{1 - z_1^2}\right)$, as required, but otherwise it is $\pm\pi - \arcsin\left(z_1\sqrt{1 - z_2^2} + z_2\sqrt{1 - z_1^2}\right)$.

It is also possible to state correct relations between the inverse trigonometric functions, as in [9]:

$$\arcsin z = \arctan\frac{z}{\sqrt{1 - z^2}} + \pi\mathcal{K}(-\ln(1 + z)) - \pi\mathcal{K}(-\ln(1 - z)). \tag{20}$$

No really new issues arise when looking at the other inverse trigonometric functions, or at the inverse hyperbolic functions.

5   The Case for `ATAN2`

It is common to say, or at least believe, that, for real x and y,

$$\arg(x + iy) = \arctan\left(\frac{y}{x}\right), \tag{21}$$

but a moment's consideration of ranges (a tool that we have found very valuable in this area) shows that it cannot be so: the left-hand side has a range of $(-\pi, \pi]$ with the standard branch cuts, and certainly has a range of size 2π, whereas the right-hand side has a range of size π.

The fundamental problem is, of course, that considering $\frac{y}{x}$ immediately confuses $1 + i$ with $-1 - i$. This fact was well-known to the early designers of FORTRAN, who defined a two-argument function `ATAN2`, such that

$$\texttt{ATAN2}(y, x) = \arctan\left(\frac{y}{x}\right) \overset{?}{\pm} \pi. \tag{22}$$

More precisely, the correction factor is 0 when $x > 0$, $+\pi$ when $x < 0$ and $y \geq 0$, and $-\pi$ when $x, y < 0$. For completeness, one should also define what happens when $x = 0$, when the answer is $+\pi/2$ when $y > 0$ and $-\pi/2$ when $y < 0$.

---

[8] The situation with addition of arcsin is complicated: see the discussion around equation (38).

This has been added to OpenMath, as the symbol `arctan` in the `transc2` Content Dictionary. Use of this enables us to rescue the incorrect equation [2, 6.1.24] $\arg \Gamma(z + 1) = \arg \Gamma(z) + \arctan \frac{y}{x}$ (where x and y are the real and imaginary parts of z) as

$$\arg \Gamma(z + 1) \equiv \arg \Gamma(z) + \arctan(y, x) \pmod{2\pi}. \tag{23}$$

We should note the necessity to think in terms of congruences.

6 Multivalued Functions

Mathematical texts often urge us (and we have found this idea useful in [6,5]) to treat these functions as multivalued (which we will interpret as set-valued), defining, say, $\mathrm{Ln}(z) = \{y \mid \exp y = z\} = \{\mathrm{Ln}\, z + 2n\pi i \mid n \in \mathbf{Z}\}$ (therefore $\mathrm{Sqrt}(z) = \pm\sqrt{z}$) and $\mathrm{Arctan}(z) = \{y \mid \tan y = z\} = \{\arctan(z) + n\pi \mid n \in \mathbf{Z}\}$ (the notational convention of using capital letters for these set-valued functions seems helpful). It should be noted that Ln and Arctan are deceptively simple in this respect, and the true rules for the inverse trigonometric functions are [2, (4.4.10–12)]

$$\mathrm{Arcsin}(z) = \{(-1)^k \arcsin(z) + k\pi \mid k \in \mathbf{Z}\} \tag{24}$$

$$\mathrm{Arccos}(z) = \{\pm \arccos(z) + 2k\pi \mid k \in \mathbf{Z}\} \tag{25}$$

$$\mathrm{Arctan}(z) = \{\arctan(z) + k\pi \mid k \in \mathbf{Z}\} \tag{26}$$

$$\mathrm{Arccot}(z) = \{\mathrm{arccot}(z) + k\pi \mid k \in \mathbf{Z}\} \tag{27}$$

$$\mathrm{Arcsec}(z) = \{\pm \mathrm{arcsec}(z) + 2k\pi \mid k \in \mathbf{Z}\} \tag{28}$$

$$\mathrm{Arccsc}(z) = \{(-1)^k \mathrm{arccsc}(z) + k\pi \mid k \in \mathbf{Z}\} \tag{29}$$

$$\tag{30}$$

where we have changed to our set-theoretic notation, and added the last three equations, which are clearly implied by the first three.

[2, (4.4.26–31)] give equivalent multivalued expressions in terms of Ln, as in table 6 (we have preserved their notation). To get the correct indeterminacy from equation (24), it is in fact necessary to interpret $z^{\frac{1}{2}}$ as $\mathrm{Sqrt}(z)$ throughout this table. The range restrictions are in fact unnecessary (as proved in [12]), and it is possible (and consistent with the decisions in the univariate case) to accept these as definitions.

One might think that the move to multivalued functions was a simplification. Indeed many statements that needed caveats (unwinding numbers, exceptional cases) before are now unconditionally true: we give a few examples below, where, for example, $\mathrm{Ln}(z_1) + \mathrm{Ln}(z_2)$ is to be interpreted as $\{x + y \mid x \in \mathrm{Ln}(z_1) \wedge y \in \mathrm{Ln}(z_2)\}$.

$$\mathrm{Sqrt}(z_1)\,\mathrm{Sqrt}(z_2) = \mathrm{Sqrt}(z_1 z_2)$$
$$\mathrm{Ln}(z_1) + \mathrm{Ln}(z_2) = \mathrm{Ln}(z_1 z_2)$$

Table 1. Multivalued functions in terms of Ln

(4.4.26) $\operatorname{Arcsin} x = -i\operatorname{Ln}\left[(1-x^2)^{\frac{1}{2}} + ix\right] \quad x^2 \le 1$

(4.4.27) $\operatorname{Arccos} x = -i\operatorname{Ln}\left[x + i(1-x^2)^{\frac{1}{2}}\right] \quad x^2 \le 1$

(4.4.28) $\operatorname{Arctan} x = \frac{i}{2}\operatorname{Ln}\frac{1-ix}{1+ix} = \frac{i}{2}\operatorname{Ln}\frac{i+x}{i-x} \quad x$ real

(4.4.29) $\operatorname{Arccsc} x = -i\operatorname{Ln}\left[\frac{(x^2-1)^{\frac{1}{2}}+i}{x}\right] \quad x^2 \ge 1$

(4.4.30) $\operatorname{Arcsec} x = -i\operatorname{Ln}\left[1 + i\frac{(x^2-1)^{\frac{1}{2}}}{x}\right] \quad x^2 \ge 1$

(4.4.31) $\operatorname{Arccot} x = \frac{i}{2}\operatorname{Ln}\frac{ix+1}{ix-2} = \frac{i}{2}\operatorname{Ln}\frac{x-i}{x+1} \quad x$ real

$$\operatorname{Ln}(\bar{z}) = \overline{\operatorname{Ln} z}$$
$$\operatorname{Arcsin}(\bar{z}) = \overline{\operatorname{Arcsin} z}.$$

However, all is not perfect. Equation (20), which needed caveats (but only on the branch cuts), now becomes the strict containment

$$\operatorname{Arcsin} z \subset \operatorname{Arctan}\frac{z}{\operatorname{Sqrt}(1-z^2)}, \tag{31}$$

and the true identity is

$$\operatorname{Arcsin} z \cup \operatorname{Arcsin}(-z) = \operatorname{Arctan}\frac{z}{\operatorname{Sqrt}(1-z^2)}. \tag{32}$$

Note that it is not true that $\operatorname{Arcsin} z = \operatorname{Arctan}\frac{z}{\sqrt{1-z^2}}$: the right=hand side has values alternately in $\operatorname{Arcsin} z$ and $\operatorname{Arcsin}(-z)$, and misses half the values in each.

6.1   Addition Laws

[2] quotes several addition laws for the multivalued inverse trigonometric functions. We give below (4.4.32–4).

$$\operatorname{Arcsin}(z_1) \pm \operatorname{Arcsin}(z_2) = \operatorname{Arcsin}\left(z_1\sqrt{1-z_2^2} \pm z_2\sqrt{1-z_1^2}\right). \tag{33}$$

$$\operatorname{Arccos}(z_1) \pm \operatorname{Arccos}(z_2) = \operatorname{Arccos}\left(z_1 z_2 \mp \sqrt{(1-z_1^2)(1-z_2^2)}\right). \tag{34}$$

$$\operatorname{Arctan}(z_1) \pm \operatorname{Arctan}(z_2) = \operatorname{Arctan}\left(\frac{z_1 \pm z_2}{1 \mp z_1 z_2}\right). \tag{35}$$

Equation (35) is, as the layout suggests, shorthand for the two equations

$$\operatorname{Arctan}(z_1) + \operatorname{Arctan}(z_2) = \operatorname{Arctan}\left(\frac{z_1 + z_2}{1 - z_1 z_2}\right) \tag{36}$$

and

$$\text{Arctan}(z_1) - \text{Arctan}(z_2) = \text{Arctan}\left(\frac{z_1 - z_2}{1 + z_1 z_2}\right). \qquad (37)$$

It would be tempting to think the same of equation (34), but in fact $\text{Arccos}(x) = -\text{Arccos}(x)$, so the $\pm$ on the left-hand side is spurious. Modulo 2π, each of $\text{Arccos}(z_1)$ and $\text{Arccos}(z_2)$ has two values, so the left-hand side has, generically, four values modulo 2π. Therefore we seem to need (see the proof in [12]) both values of $\mp$, and this is indeed true. The equation could also be written as

$$\text{Arccos}(z_1) + \text{Arccos}(z_2) = \text{Arccos}\left(z_1 z_2 + \text{Sqrt}\left((1 - z_1^2)(1 - z_2^2)\right)\right).$$

When it comes to equation (33), the situation is more complicated, but in fact it is possible to prove (see [12]) that any containment of the form $\text{Arcsin}(z_1) + \text{Arcsin}(z_2) \subset \text{Arcsin}(A)$ must also have the property that $\text{Arcsin}(z_1) - \text{Arcsin}(z_2) \subset \text{Arcsin}(A)$. So the equation should be read as

$$\text{Arcsin}(z_1) \pm \text{Arcsin}(z_2) = \text{Arcsin}\left(z_1 \text{Sqrt}(1 - z_2^2) + z_2 \text{Sqrt}(1 - z_1^2)\right), \qquad (38)$$

with each side taking on eight values modulo 2π (counting special cases like $\text{Arcsin}(1)$ as a "double root").

It is unfortunate that the desire to save space led the compilers of [2] to compress equations (36) and (37) into equation (35), since the $\pm$ notation here actually has a completely different meaning from its use in the adjacent equations (33) and (34). For completeness, let us say that in [2, (4.4.35)] —

$$\text{Arcsin}\, z_1 \pm \text{Arccos}\, z_2 = \text{Arcsin}\left(z_1 z_2 \pm \sqrt{(1 - z_1^2)(1 - z_2^2)}\right)$$

$$= \text{Arccos}\left(z_2 \sqrt{1 - z_1^2} \mp z_1 \sqrt{1 - z_2^2}\right)$$

the convention is as in (4.4.32), i.e. the equation cannot be split and $\sqrt{w}$ means $\text{Sqrt}(w)$, whereas in [2, (4.4.36)] —

$$\text{Arctan}\, z_1 \pm \text{Arccot}\, z_2 = \text{Arctan}\left(\frac{z_1 z_2 \pm 1}{z_2 \mp z_1}\right)$$

$$= \text{Arccot}\left(\frac{z_2 \mp z_1}{z_1 z_2 \pm 1}\right)$$

the convention is as in (4.4.34), i.e. the equation can be split.

7   Couthness

[9] introduced this concept. If h is any hyperbolic function, and t the corresponding trigonometric function, we have a relation

$$t(z) = ch(iz) \text{ where } c = \begin{cases} 1 & \cos, \sec \\ i & \cot, \text{cosec} \\ -i & \sin, \tan \end{cases}. \qquad (39)$$

From this it follows *formally* that

$$h^{-1}\left(\frac{1}{c}z'\right) = it^{-1}(z').$$ (40)

Definition 1. *A choice of branch cuts for h^{-1} and t^{-1} is said to be a* couth *pair of choices if equation (40) holds except possibly at finitely many points.*

[9] show that, with their definitions (the definitions of [2] with the values on the branch cuts prescribed) all pairs were couth except for:

arccos/arccosh Here equation (40) only holds on the upper half-plane (including the real axis for $\Re z \leq 1$);

arcsec/arcsech Here equation (40) only holds on the lower half-plane (including the real axis for $\Re z > 1$).

However, [2, (4.4.20–25)] show that all pairs are couth in the multivalued case (where equation (40) is interpreted as equality of sets).

8 Conclusion

This paper has, as is perhaps inevitable at this stage of Mathematical Knowledge Management, posed more questions than it answers. For convenience, we recapitulate them here.

1. Should we codify a branch cut, e.g. for ln as a direct subset of **C**, or via a specification such as equation (5).
2. If the former, what formal semantics can we attach to the phrase "branch cut"? Can one do this in a way independent of the specification of **C** and **C** $\rightarrow$ **C**?
3. What should be the correct encoding of false equations such as equation (1): unwinding numbers, conditional or multivalued? How does one cope with equations such as (7) and (8) in the conditional formalism — aren't we just rewriting the unwinding number formalism? Conversely, equation (16) is distinctly simpler than equation (15), and equation (19) currently has no unwinding equivalent. Should a Mathematical Knowledge Management system (in this area) have to support more than one such encoding?
4. How do we support the restriction of these functions to (partial) functions **R** $\rightarrow$ **R**? In this case *most* of the unwinding number terms or conditions drop out. It is harder to see how the multivalued formalism supports this restriction.
   The obvious case where some caveat is still necessary is $\sqrt{z^2}\stackrel{?}{=}z$, where the formalisms might be:
   $$z \geq 0 \Rightarrow \sqrt{z^2} = z;$$
   $$\sqrt{z^2} = (-1)^{\mathcal{K}(2\ln z)}z.$$
   The second has the disadvantage of still introducing complex numbers, via $\ln z$ when $z < 0$, though it could clearly be massaged into $\sqrt{z^2} = (\text{sign } z)z$.

5. It appears that, contrary to popular belief, the multivalued semantics are not simply a tidier version of the univalued (branch cut) semantics: contrast equation (20) and its conditional equivalent

$$(z \notin \mathbf{R}) \vee (z^2 \leq 1) \Rightarrow \arcsin z = \arctan \frac{z}{\sqrt{1 - z^2}}$$

with equation (32). Does this mean that we need two separate Mathematical Knowledge Repositories for the two cases?

6. Can a Mathematical Knowledge Repository for these facts (either case, or both cases) be simply a database, or must it be much more intelligent, possibly incorporating ideas along the lines outlined in [5].

We also deduce the following differences between the "Abramowitz & Stegun" (A+S) era and the MKM era.

- In the A+S era, it was not necessary to specify the values of the functions on branch cuts: numerical analysts for the most part did not care (but see [15]) since the branch cuts were of measure zero, and the intelligent reader could choose the adherence most suitable to the problem. In the MKM era, both computer algebra systems and theorem provers need to know correctly what the answer is. For interoperability, they must agree on what the answers is — see the examples in [9].
- In the A+S era, it was acceptable (maybe only just) to use the $\pm$ notation to mean two different things: in the MKM era it is not, and the notation should only be used (if at all) with $A \pm B$ being shorthand for $\{a + b, a - b\}$ (or, in the set-valued case $(A + B) \cup (A - B)$).
- A+S was ambivalent about whether it was talking about $\mathbf{C} \to \mathbf{C}$ or (partial) $\mathbf{R} \to \mathbf{R}$. Many of the formulae are stated with (unnecessary) restrictions to the $\mathbf{R}$ case — see equation (13) and [2, 4.4.28] relating Arctan to Ln, which restricts z to be real.
- In the A+S era "everyone knew" what a branch cut was. To the best of the author's knowledge, no computer algebra system or theorem prover does.

It is hoped that these thoughts, limited as they are to one chapter of one book, will stimulate debate about the difficulties of managing this sort of mathematical knowledge.

References

1. Abbott,J.A., Díaz,A. & Sutor,R.S, OpenMath: A Protocol for the Exchange of Mathematical Information. SIGSAM Bulletin **30** (1996) 1 pp. 21–24.
2. Abramowitz,M. & Stegun,I., Handbook of Mathematical Functions with Formulas, Graphs, and Mathematical Tables. US Government Printing Office, 1964. 10th Printing December 1972.
3. Aslaksen,H., Can your computer do complex analysis?. In: *Computer Algebra Systems: A Practical Guide* (M. Wester ed.), John Wiley, 1999.
   http://www.math.nus.edu.sg/aslaksen/helmerpub.shtml.

4. Beeson,M. & Wiedijk,F., The Meaning of Infinity in Calculus and Computer Algebra Systems. Artificial Intelligence, Automated Reasoning, and Symbolic Computation (ed. J. Calmet *et al.*), Springer Lecture Notes in Artificial Intelligence 2385, Springer-Verlag, 2002, pp. 246–258.

5. Bradford,R.J. & Davenport,J.H., Towards Better Simplification of Elementary Functions. Proc. ISSAC 2002 (ed. T. Mora), ACM Press, New York, 2002, pp. 15–22.

6. Bradford,R.J., Corless,R.M., Davenport,J.H., Jeffrey,D.J. & Watt,S.M., Reasoning about the Elementary Functions of Complex Analysis. *Annals of Mathematics and Artificial Intelligence* **36** (2002) pp. 303–318.

7. de Bruijn,N., The Mathematical Vernacular, a language for mathematics with type sets. Proc. Workshop on Programming Logic, Chalmers U., May 1987.

8. Corless,R.M., Davenport,J.H., Jeffrey,D.J., Litt,G. & Watt,S.M., Reasoning about the Elementary Functions of Complex Analysis. Artificial Intelligence and Symbolic Computation (ed. John A. Campbell & Eugenio Roanes-Lozano), Springer Lecture Notes in Artificial Intelligence Vol. 1930, Springer-Verlag 2001, pp. 115–126.

9. Corless,R.M., Davenport,J.H., Jeffrey,D.J. & Watt,S.M., "According to Abramowitz and Stegun". *SIGSAM Bulletin* **34** (2000) 2, pp. 58–65.

10. Corless,R.M. & Jeffrey,D.J., The Unwinding Number. SIGSAM Bulletin **30** (1996) 2, pp. 28–35.

11. Davenport,J.H., Table Errata — Abramowitz & Stegun. To appear in *Math. Comp.*

12. Davenport,J.H., "According to Abramowitz and Stegun" II. OpenMath Thematic Network Deliverable ???, 2002. http://www.monet.nag.co.uk/???/

13. Dewar,M.C., OpenMath: An Overview. *ACM SIGSAM Bulletin* **34** (2000) 2 pp. 2-5.

14. IEEE Standard 754 for Binary Floating-Point Arithmetic. IEEE Inc., 1985.

15. Kahan,W., Branch Cuts for Complex Elementary Functions. *The State of Art in Numerical Analysis* (ed. A. Iserles & M.J.D. Powell), Clarendon Press, Oxford, 1987, pp. 165–211.

16. Rich,A.D. and Jeffrey,D.J., Function evaluation on branch cuts. SIGSAM Bulletin 116(1996).

17. Stoutemyer,D., Crimes and Misdemeanors in the Computer Algebra Trade. *Notices AMS* **38** (1991) pp. 779–785.

From Proof-Assistants to Distributed Libraries of Mathematics: Tips and Pitfalls

Claudio Sacerdoti Coen

Department of Computer Science
Mura Anteo Zamboni 7, 40127 Bologna, ITALY.
sacerdot@cs.unibo.it

Abstract. When we try to extract to an open format formal mathematical knowledge from libraries of already existing proof-assistants, we must face several problems and make important design decisions. This paper is based on our experiences on the exportation to XML of the theories developed in Coq and NuPRL: we try to collect a set of (hopefully useful) suggestions to pave the way to other teams willing to attempt the same operation.

1 Introduction

The formalization of interesting and useful fragments of mathematics and computer science requires the development of a lot of elementary theories. In particular, we recently got more and more evidence of the fact that complex chapters of mathematics may be required even to justify the correctness of simple algorithms [8,1].

Current tools for the development of formal mathematical knowledge, such as theorem-provers and proof-assistants, have been successful in the last decade only in encoding small parts of mathematics and computer science, starting from the elementary ones. The lack of sharing between the different systems resulted in many theories formalized again and again, each time using a different tool. Moreover, since all of these systems are application-centric and use proprietary formats, the formalized knowledge is not even accessible, if not by means of the system itself. Finally, the tasks of visual rendering mathematical libraries and making them searchable and reusable are complex issues that are usually under-developed in these tools.

Several groups and projects[1] tried to improve the situation, providing two general and largely orthogonal solutions.

The first solution consists of *making the systems communicate* to each other, developing some kind of lingua franca (e.g. OpenMath [12]) and standardizing software interfaces (e.g. the MBase mathematical software bus [9]). This is the solution that less impacts the systems, since it is enough to provide a module

---

[1] The Calculemus Network, the MKM Network, the OpenMath Society, the HELM Project, the MoWGLI Project, the FDL Project, the MathWeb Project and many others.

A. Asperti, B. Buchberger, J.H. Davenport (Eds.): MKM 2003, LNCS 2594, pp. 30–44, 2003.

(called Content Dictionary in OpenMath) to query other systems and answer to queries, mapping the lingua franca to their logical dialect. This solution works well for computer algebra systems and first order theorem provers, which are all based on a very similar logic and which are quite flexible on the notion of correct mathematical results: since the logic is more or less the same, a result proved by another system can be reasonably considered correct. Fewer applications have been proposed for making proof-assistants cooperating. The reason is that traditionally these systems do not attempt proof-search for undecidable theories and thus are allowed to exploit more complex logical frameworks (Higher Order Logics, Pure Type Systems, Calculus of Constructions, Martin-Löf Constructive Type Theory) with more emphasis on the foundational issues and on the notion of correctness[2]. Since every proof-assistant adopts a different logic, proofs from other systems can not be reused nor trusted, unless encodings of one logic into the others are provided[3]. Finally, note that this solution is successful in avoiding proving the same things again or implementing the same decision procedures, but does not address other problems, in particular those related to the publishing and rendering of the developed theories.

The second solution, pursued by other projects [2,4], is to develop standard formats for long-term storing of the formal mathematical knowledge and to use them to build distributed libraries of documents. In a previous paper [3] we explained why XML-based formats are a promising choice. Once the standard formats have been defined, tools can be independently provided to manage, render, index, transform, create, check and data mine the documents in the library. Since we must face the multilingual environment of the many logical frameworks, we can not have just one format, but we need at least one for each logic. Nevertheless, those operations which are independent of the specific logic can still be implemented once and applied to every document, either directly or after a transformation that maps the logic level format to a Content Level one (such as MathML Content or OpenMath); the transformation prunes out the encoding details of the rigorous mathematical notion into the logical framework, obtaining an informal representation that is more suitable for searching, rendering and exchange (by means of Phrasebooks) with other systems that are based on a different logic. Instead, logic dependent tools such as proof-checkers, program extractors and decision procedures can work directly on the format that corresponds to their logic, without any need of importing or exporting the documents.

---

[2] Axioms are (almost) never introduced in these systems, so that the consistency of the theory developed is reduced to the consistency of the logical system. As a consequence, everything must be proved, here comprising very primitive notions such as the fact that 0 is different from 1.

[3] Note that a notion P in the encoding of a system A in a system B is generally different from the same notion P given directly in system B. For example, 0 is defined in set theory as the empty set and in type theory as a constructor of an inductive type. Given that set theory can be encoded into type theory, we can define a new 0 as the (encoded) empty set; still we will not be able to use it in place of the original 0, since its type (and properties) are different

There is an alternative to defining a new XML format for each logic: Open-Math, being an extensible format, can be used to encode any logical system, simply defining in OpenMath a symbol for each constructor of the logic. In this way, though, we do not get any real benefit, since there is no sharing of constructors with other logics; so logic dependent tools will have Phrasebooks defined on disjoint subsets of OpenMath and will still not be able to share information, unless a transformation to a Content Level is provided. The only tools that could be shared are parsers and tools to manage the distribution of the documents, but with no real benefit. Parsing is not made easier with respect to standard XML parsing, since the logic level tools will have to interpret the parsed structures to recognize the constructs of the logic and reject other constructs; ad-hoc Document Type Definitions (DTDs), instead, at least relieve the system from this rejection phase. Distribution tools do not get any benefit either, since they do not need to interpret the data[4]. On the contrary, at the Content Level Open-Math provides already developed libraries for the management and exchange of the information, and may enhance comunication by means of modularly designed Phrasebooks.

While the first solution does not require big modifications in the existent tools, the second approach is much more invasive (but also much more promising), because it implies for the existent tools the change from their application-centric architecture to the document-centric one. Since this is a slow process, the suggested shortcut is to provide modules for the existent systems to extract the mathematical knowledge from their libraries and save it in the new standard format. This way we achieve two different and equally important goals: we can start developing the new tools around an already existent library of interesting documents and we preserve the already formalized content, avoiding redoing the formalization and, in the long term, improving the way it can be accessed and exploited.

In 1999 we wrote for project HELM (Hypertextual Electronic Library of Mathematics) a DTD for a variant of the Calculus of (Co)Inductive Constructions (CIC), which is the logical system of the Coq[5] proof-assistant, and an exportation module from the Coq compiled library files (".vo" files) to our XML format. Several tools were then developed to work on the exported library: a distribution and publishing system; stylesheets to render proofs and definitions in MathML Presentation or HTML; dependency analyzers and renderers for studying the relations among items of the library; a query language based on automatically generated metadata to summarize relevant information extracted from the statements of the theorems; a brand new stand-alone incremental proof-checker; a prototype of a proof-engine. Of the previous tools, all but the last two are logic independent (which implies DTD-independent). All the other tools can be reused on any other DTD used to encode any other logical framework. The

---

[4] They may inspect metadata, but metadata are provided using the standard RDF format and are completely logic independent.

[5] Another proof-assistant based on a different variant of the same calculus is Lego.

development of our tools helped us in identifying wrong decisions in the design of the DTD and in the implementation of the export module.

In 2002 we became members of the European IST Project MoWGLI (Math on the Web, Get it by Logics and Interfaces), which extends the goal of the HELM project and adds new requirements. Sustained by our previous experience, we are redefining the DTD for the Calculus of (Co)Inductive Constructions and we are reimplementing the exportation module, starting from the first version and adding several new features.

In the meantime in the HELM project we are now trying to export also the library of the NuPRL system, which is based on a variant of the Martin-Löf constructive type theory.

In the rest of the paper we will try to outline the principles that we have learnt in the last three years and that we are now trying to apply to our new implementations of exporting modules.

2   Tips and Pitfalls

The main self-assesment criteria for an exportation module is that the exported information must be complete, general enough and unbiased, in the sense that it must not force some design choices that we would better avoid. The risk is to later find out to be unable to develop the expected tools.

Thus we must face at least the following issues: What information should be exported? What is the granularity of the generated files? How is the information described? (i.e. how should a DTD be designed)

2.1   What Classes of Information Should Be Exported?

Catalogue the information according to its use.
When the information is required for more than one task, factorize.

Before writing the DTDs, catalogue all the information that is available inside the system and all the information you are interested in (which may or may not be already available). Since different tools may require different subsets of the whole knowledge and since you are likely to change the DTDs and the documents often, insulating the tools from changes in other parts of the library is almost mandatory. It is not unusual to find out later that the grain was not fine enough. Example: not every operation that is applied to a theorem requires both the statement and the proof. Finding out which lemma can be applied, for instance, just requires the statement or, even better, the list of its hypotheses and the conclusion.

Sometimes some data is required to perform more than one task. In that case it is better to factorize, even if the factorized information does not seem at once to have any special meaning.

We have identified the following classes of information in the library of the Coq system:

- Definitions, inductive definitions and proof-objects as sets of lambda-terms, according to the Curry-Howard isomorphism and the theory of CIC. Even for definitions, this is not what the user entered to the system, but the result of complex post-processing, typing and refining rules implemented in Coq. This is what is actually *proof-checked*, by type-checking the lambda-terms and testing the convertibility of their inferred type (in case of a proof, what the theorem really proves) with the expected type (the statement of the theorem). Theorems are further refined (see below).
- Statements of the theorems. They are useful not only for proof-checking. First of all the user can be interested just in the statement of a theorem, that can be parsed and rendered independently of its proof. Secondly they are the only information required to answer several kind of logic-dependent queries:
  1. which lemma can be applied in this situation?
  2. which theorem concludes an instance or a generalization of something?
  3. what can we conclude from a certain set of hypotheses?

  Note that the previous queries are essentially based on the notion of unification, which is an expensive operation when performed on very large sets of candidates (that must also be parsed and loaded into memory to apply unification).

  Statements of the theorems can be further divided into hypotheses and conclusion. This is not a trivial task, since this information is encoded in a lambda-term and because the constructors of the logical framework (Π-abstractions) are overloaded in Coq to mean either a dependent product, a non dependent product (function space), a logical implication or a universal quantification. Other logical frameworks as the one of NuPRL, instead, can have several primitive or derived constructors for introducing hypotheses. Thus, to implement the logic independent queries in a general way, the separation must be performed in advance (either statically or dynamically, when needed).
- Logic independent information extracted from a statement. For example, the single notion of occurrence of a definition in a statement is useful to answer interesting queries:
  - which theorem states some property of an operator?
  - which theorem *may* be applied in a certain situation? Note that if we want to prove some fact about the multiplication of two real numbers, we are interested only on those statements where that multiplication occurs and no other uninteresting definition (let's say the "**append**" function of two lists) occurs. So we can effectively use this information to quickly filter out those theorems whose application will never succeed. Note that this filtering operation is logic independent (then it can be provided once and for all for every system), it is easily implemented using a standard relational or XML-based data-base and can be extremely more efficient than trying one at a time the applications, which usually involve higher-order unification of two expressions.

- Proofs. Their size is usually orders of magnitude bigger than the size of their statements. They are never rendered alone, but are an interesting part of the library for data-mining (for example to recognize similar proofs or to understand the complexity of some proofs). An usual operations on proofs that do not require additional information is proof improvement: a group of inference steps, possibly automatically found by the system, may be replaced with a shorter proof, usually human provided.

- Logic independent information extracted from a proof. As for the case of statements, the most interesting notion is that of occurrence. Given the list of occurrences it is easy to answer the following queries:

  • which proofs depend, directly or indirectly, on a given lemma or axiom?
  • which axioms does a proof depend on, either directly or indirectly?
  • which part of the library will be affected if some definition or axiom is changed?
  • what should I learn to be able to understand the following theorem?

- System dependent information related to a proof or definition. For example, in Coq and other systems there exists a notion of implicit arguments, which are those arguments that can be automatically inferred by the system if they are not provided. For example, when writing $x = \pi$ it is clear that the monomorfic equality of type $\forall T.T \to T \to$ **Prop** is applied to the set of real numbers. This information is not necessary to proof-check a document, but may be useful to render it: information that can be inferred by a system can often be inferred by the human being and is usually omitted in an informal presentation. Finally, note that implicit arguments are really system dependent, since a more powerful system may be able to infer more information and thus consider more arguments to be implicit.

- Redundant information related to a proof or definition. A typical example is the opacity of a constant. Opaque constants are abstract data types, whose exact definition can not be inspected. Proofs in proof-irrelevant systems are always opaque. In those systems that are not proof-irrelevant, such as Coq, all the constants may be considered transparent for the sake of proof-checking. This means that all constants may be expanded during proof-checking. Often, though, it is not necessary to expand every constant for proof-checking and knowing in advance which constants may not be expanded can make the system much more performant. This information is essentially logical redundant, since there is an easy algorithm to make it explicit: try type-checking without any expansion and, in case of failure, backtrack and expand. Of course the computational complexity of this algorithm is, in the worst case, exponential in the number of constants that occur in the theorem being typed.

Another even more interesting example of redundant information is the types of the sub-expressions of one proof, that corresponds to the conclusions of the subproofs. This information is completely logical redundant (if type-inference is decidable), but it is essential to render the term in a pseudo-natural language [6].

Since extracting the implicit information inside the system is easy, but it may be very difficult, time consuming or hardly possible to do with an external tool, the advice is to

> Make implicit information explicit.

- Metadata related to definitions and theorems. These are other logic independent information such as author name or the version of the system the proof was developed in that can be useful to implement other kind of queries. Metadata can range from simple to very complex ones, as those needed in educational systems [10]. Usually, though, they are not provided inside the system or they are just given as unstructured comments.
- A system dependent history of the operations that lead to the creation of a definition or theorem. For example, in the case of Coq, we can export the proof-tree, which is the internal structured representation of the list of tactics (and their arguments) used to prove one theorem. This information may be useful both for rendering purposes and to replay the proof inside the system, in case we need to modify it.
- Comments. Comments provided by the users are an extremely valuable form of documentation. The problem is that they are often discharged during the lexical analysis of the input to the system; So they are likely to be unavailable inside the system itself.
- Parsing and pretty-printing rules. The existent systems only perform pretty-printing to ASCII notation, while we are interested in more advanced visual rendering. Thus we do not have any use for this information, that will not be exported.
- Tactics and decision procedures. It would be extremely interesting to put the code of tactics and decision procedures into the library. In this way it would become possible to replay the construction of a proof independently from the system that generated it and from its version. Indeed a problem we face is that every time some detail of the implementation of Coq changes, the same script produces a new slightly different proof. This represents a problem from the point of view of proof-engineers, that must continuously update the proofs every time a definition changes.

  To have the tactics into the library we need first to formalize the language (and the libraries) in which they are implemented and this is surely a complex task.

Once the information classes have been identified, it is time to start developing a syntax (a DTD in the case of XML) for them. At this stage it is also important to understand what are the relations between the data belonging to the different classes. Those relations must be made explicit. Thus the third advice is

> Develop one DTD for each class you are interested in.
> Make the links between different instances explicit in the DTD.
> Locate the atomic components.

Our initial expectation was to be able to heavily link at a very fine grained level the information collected in the documents belonging to the different classes. For example, it is reasonable to link:

- every type of a sub-term of a proof (inner-type) to that sub-term;
- every node of the proof-tree (that corresponds to a user or system issued tactic) to the sub-term it generated.

Still it is often the case that there is no real correspondence between those notions inside the system. For example it may happen that a tactic generates a term that is not present in the final proof, because a subsequent tactic modified or discarded it. Another possibility is that the whole proof can be generated using dirty tactics that create detours or lot of redexes that are later simplified during a clean-up stage. We consider this behaviour of the system to be a debatable design decision, since it introduces a gap between the proof that the user wants to describe and the proof that is produced at the end. This may be especially annoying when the user is really interested in the generated proof-object and he is not given any effective tool to inspect it.

In these cases, instead of simply avoiding to provide that linking information, it is better to pinpoint to the developers of the system the problematic operations to be modified. In the worst case it may be necessary to create a branch in the system development to obtain a slightly modified tool that is used to produce the library that will be eventually exported.

The atomic components that must be located belong to the same information classes and represent the minimal referentiable units that have a precise meaning when considered as stand-alone documents[6]. Sometimes this notion is quite fuzzy. For example, what are the atomic components of a block of mutually defined functions? We may choose that the only atomic component is the whole block, and the functions are simply subparts that may be referenced starting from the whole block (using XPaths, in the case of XML). Or we may consider each function to be atomic and make it refers to the other mutual defined functions via links. Note that some operations (rendering, proof-checking) require the whole block, while others (extraction of metadata used to answer queries) operate on one function at a time. Thus, according to our first suggestion, we should split the functions.

2.2   How Should Files Be Organized?

The need to clearly identify the atomic components, addressed in the previous sections, is related to our next suggestion:

> Do not mix information about several components or belonging to several classes: one file for each class and for each component.

---

[6] Of course the documents may have links to other documents that must be considered when determining their meaning. What is important here is that if some kind of context is requested, it must be explicitly stated by means of explicit links.

Assembling together information from different classes is perhaps the worst pitfall, for several reasons:

- The set of operations we want to perform on the data is not a fixed one. For example, we can implement several indexing tools to be able to answer to different queries. Since every tool requires and produces new information that should be put into the library, it is fundamental to be able to define new file formats (DTDs) and change existent ones. If several information classes are assembled together, this means that we will have to modify all the already developed tools to handle (usually to ignore!) the new or modified syntax. Moreover we will have to regenerate huge amounts of files distributed over the network (we are supposed to work with distributed libraries of formal mathematical knowledge).
- The amount of information we want to handle is large. For example, just the proof objects of the Coq libraries and their inner-types, once saved as compressed XML files, require about 260Mb. It is not unusual to have single theorems that, exported in XML and compressed, require 24Kb. Thus parsing (or even lexically analyze) these files is an expensive operation. Of course we want to avoid as much as possible wasting time in parsing parts of a file we are not interested in.
- It is reasonable, at least for some applications, to use a data-base to hold the exported information. In that case the XML representation is used just as an exchange format and mixing several data in the same file is not a problem, since the information will be disgregated inside the data base. This approach, though, limits a priori the possibility of implementing other tools that, instead of connecting themselves to the database, will just work on the original files. Moreover, the database approach is surely heavier from the point of view of Web-publishing, where the most successful model is that of Web pages publishing, which is simpler and less demanding.

So far we have given suggestions on how to decide which classes of information should be exported and how files should be organized. We will now go back to the problem of designing the DTD and focus on the main difficulties that must be faced.

2.3   What Theory Should Be Described?

Describe, as much as possible, the theory and not the implementation.
Do not be afraid of mapping concepts from the implementation back to the theory.
Make the theory explicit.
Introduce new theories to justify the implementation.

There may be a big gap between the theory a system is based on and the way it is implemented. Proof-assistants are implemented starting from a well-known theory (e.g. Martin-Löf Constructive Type Theory for NuPRL; the Calculus of

Constructions for Coq). Later on, extensions to the theory are implemented and possibly proved to be correct. For example, both in the case of NuPRL and HOL, logic constructions that are derived in the original theory are made primitive in the implementation for efficiency reasons. In NuPRL this is done, for example, for the integer numbers and the usual arithmetic operations on them.

What usually happens is that after some years it is difficult to reconstruct a unified theory of all the extensions provided, while many other changes are completely undocumented. This happens not only in the kernel of the systems, which is responsible of checking that inference steps are well-applied, but also in external layers that rely on techniques (e.g. higher order unification) that are well-known in the literature and that requires extensions to the representation of the proofs (e.g. metavariables that are typed holes in a proof-term).

When exporting from Coq the first time, we decided to be quite close to the implementation, even if some parts of what we exported were partially not understood. When we developed our proof-checker first and our proof-assistant later, we realized that we had no clear foundation for the theory we were implementing. Moreover, having represented not the terms of the theory but their encoding in the structures of Coq, the implementation of many operations were forcedly isomorphic to what was done in Coq, even if we wanted to experiment with other solutions. Let's see in detail a couple of examples.

Metavariables and Existential Variables. According to the Curry-Howard isomorphism, a correct proof can be seen as a well-typed lambda-term in some lambda-calculus. Thus an incomplete, partially correct proof must be a well-typed lambda-term with holes therein. To extend the notion of well-typing to terms with holes, a hole (called *metavariable* in the literature [11]) can not simply be a missing term, but must be associated to a sequent. So a metavariable has a given type and a typed context.

The two main operations on metavariables are instantiation and restriction. A metavariable can only be instantiated with a term that is closed w.r.t. the metavariable context and that has the expected type in that context. Instantiating a metavariable is an heavy operation, since it requires a linear visit of the whole proof-term. The other operation, restriction, deletes some hypotheses from the metavariable context. It is required, for example, to perform unification: to unify two metavariables the first step is to restrict both of them so that their contexts become equal (possibly up to a decidable convertibility relation).

How can those operations be implemented efficiently? One possibility is to implement restriction using instantiation: every time a metavariable should be restricted, a new metavariable of the right shape is generated and used to instantiate the old one. Since restriction occurs quite often, this implies that an efficient way to perform instantiation must be designed. This is the approach of Coq: restriction is reduced to instantiation and instantiation is not performed explicitly, but delayed using a new environment that maps every instantiated metavariable to the term used to instantiate it.

A completely different approach implements restriction explicitly, making each hypothesis in the sequent optional, so that it is possible to remove one hypothesis keeping trace of the fact that it was removed[7]. Nothing is done to speed up instantiation, that becomes a seldom required operation.

The two possibilities requires different data-structures and representation of metavariables. Moreover, for technical reasons, in Coq there is a further distinction between full-fledged metavariables (called existential variables and use to represent holes in the type of other metavariables) and restricted metavariables that are used to represent just the open goals.

The first time we exported from Coq, we did not change the Coq representation. Later on, when implementing our own proof-assistant, we found out that our choice forced us to a treatment of metavariables similar to the one of Coq. Moreover the distinction between the two kinds of metavariables did not allow any progress in the proof[8]. So we defined a different internal representation of metavariables, following the second solution above, and we decided to change the DTD.

Now, which is the right format for describing metavariables in the library? The only reasonable choice is to stick to the theory, where hypotheses in a metavariable context are not optional and there is no environment to delay instantiation. It is a responsibility of the systems to map back and forth between this standard and well-understood format and their internal encoding.

Sections, Variables, and Discharging. In the syntax of Coq it is possible to abstract a group of definitions and theorems with respect to some assumptions. Example (in Coq syntax):

```
Section S.
 Variable A : Prop.
 Variable B : Prop.
 Definition H1 : Prop := A /\ B.
 Theorem T1 : H1 -> A.
  Proof.
  <some proof>
  Qed.
End S.
(* Here the type of H1 and T1 are different. See code fragment below *)
Theorem T2 : True /\ True -> True.
 Proof.
  Exact (T1 True False (I,I)).
 Qed.
```

The previous fragment should be equivalent, from a logic point of view, to the following input:

---

[7] This information is required for managing explicit substitutions that are needed to allow reduction of terms with metavariables.

[8] This is not a problem in Coq since the proof-tree with holes is generated on-demand only for pretty-printing and exporting purposes. Proof-trees are instead used to describe an incomplete proof and the progress on it.

```
Definition H1 : Prop -> Prop -> Prop := [A:Prop ; B:Prop]A/\B.
Theorem T1 : (A:Prop ; B:Prop)(H1 A B) -> A.
 Proof.
 <some slightly different proof>
 Qed.
Theorem T2 : True /\ True -> True.
 Proof.
  Exact (T1 True False (I,I)).
 Qed.
```

The operation that transforms the first code fragment in the second one is called *discharging*. Discharging is implemented in an external layer of the Coq system, in such a way that the kernel of the system is given the discharged term (and the theory of the kernel of Coq does not need to be modified).

Since, for rendering purposes, we are more interested in the first fragment, we exported the undischarged form of the theorems and definitions. The problem with that representation is that the theorem T1 that is used in theorem T2 is no more equal to the one defined above: its type is different! This implies two kind of problems in developing tools:

1. While rendering T2, we would like to make the occurrence of T1 an hyperlink to its definition. What we get is misleading for the reader: the theorem T1 is shown to have a type that is not the same of its occurrence in T2.
2. To proof-check T2 we need the type of the discharged form of T1. So we are obliged to discharge T1 and this leads to serious problems: either we save the discharged form, as Coq does, and this goes against our initial choice; or we discharge the theorem on-the-fly when needed, and, being this an expensive procedure, we have to implement complex caching machineries[9].

The solution we are adopting now is simply to redesign the theory, replacing the notion of discharging with that of explicit named substitution. So, while exporting, we completely change the definition of T2 to the following one:

```
Theorem T2 : True /\ True -> True.
 Proof.
  Exact (T1[A := True ; B := False] (I,I)).
 Qed.
```

In this way we are no longer exporting the exact definition of T2 inside Coq, but something more well-behaved for our purposes and that can be mapped back, if needed, to the Coq representation. Both of the previous problems are solved with this representation, since T1 can be rendered as it is just adding the explicit substitution to the top and T2 can be type-checked without discharging T1, by taking care of the explicit substitution in the typing rules of the system.

If we stop reflecting on the real cause of the problems given by discharging, we will easily identify another severe problem that must be faced when exporting

---

[9] This is what we implemented, but the nature of the discharging operation interfered with the usual locality reference principle of caches. As a result we got very poor and unexpected performances.

the information: a proof-assistant may be implemented in an imperative way, where there is a strong notion of time and the objects of the library change from time to time. This behaviour seems utterly incompatible with the notion of mathematical library:

> A library, due to its nature, is a random access structure, both in space and time. Try to minimize the dependencies, give them the right direction, beware of "imperative" commands of the system that change already defined things. Make them immutable.

How can we face the situation in which we have to export some imperative information? For example, Coq has many other examples of this kind of information:

- A list of theorems that are considered while automatically searching a proof. Theorems can be added and removed from this list at pleasure.
- Implicit variables. Even if the system automatically chooses the variables it thinks it may infer, the user can force at any time the set of implicit variables for an object. Subsequent commands will use the new set.
- Opacity. It can be changed from transparent to opaque, to simulate abstract data types.
- Parsing and pretty-printing rules.

The situation can be described in two ways. The first one is to add the imperative command that changes the system status to the library, adding links from the command to all the other objects that are affected. The second way is to say that every command or definition of the system has an implicit dependency on the state of Coq. This dependency can be effectively make explicit. While the first solutions is easier to implement, because that information is already available in the system, interpreting a command means in practice re-running all the commands in a clone of the Coq system to build the status requested. The second solution, instead, avoids rebuilding the status. Remember that, in a distributed setting, collecting all the given commands and definitions just to update them because of imperative commands is already an unaffordable operation. Managing a distributed status may be even worse. The price to pay, of course, is that objects that change are replicated again and again for each status they may have. Our experience shows that either the requested status information is minimal or there is some problem in the formulation of the theory that should be better removed, as in the case of discharging.

3   Further Discussion and Future Work

The previous sections should make evident that there is a conflict between the two design principles *"what is expensive is worth saving"* and *"what is computable can be disregarded"*. Indeed it is often the case that the computations that can be disregarded are also the expensive ones.

We recall that the two strong motivations for the second rule are: redundant information requires consistency checks; redundant information may require unwanted parsing of unuseful data. Our suggestion is to look for a compromise preferring to avoid redundant information when possible. When this is not the case, the expensive computed information should be stored in a different file, so that we can grant at least internal consistency, i.e. consistency of the data in the same file. Internal consistency is often enough for our needs. Unwanted parsing is also minimized in this way.

Future work will consist in trying to apply our guidelines to other proof-assistants and theorem provers, with the aim of integrating them in our distributed library.

So far we have not tried to apply our guidelines to export information from a Computer Algebra System (CAS). Our feeling is that some of the issues we faced would be irrelevant for Computer Algebra Systems. For example, the theory behind these systems is quite uniform and better understood, just being some extension of a first-order setting. Determining the information classes to export should present more or less the same problems. Probably, though, other issues may get much more severe and new ones will certainly arise: CAS are more imperative than proof-assistants and theorems prover, and a notion of library is much more difficult to define. Since the power of a CAS mainly derives from the procedures it implements and not from a library of pre-computed results, exporting a description of these procedures to the library could be compelling.

Since our experience derives from a long process of committing mistakes and finding solutions, we want to conclude with a final few advices: be ready to go back on your steps; and be always creative and flexible:

Sometimes it is better to avoid good suggestions and look for compromises!

4 Conclusions

We presented a collection of simple rules-of-thumb that we learned in the previous three years from small successes and lots of mistakes. Each rule was justified with small examples; their only aim was to give a grasp on the set of problems that must be faced by a designer and implementor of an exporting procedure from a proof-assistant to a distributed library of mathematical documents.

To our knowledge, this is the first paper in the literature with this aim, even if many other exporting procedures have been developed for several other systems in the past few years [7,5].

Even if many of the suggestions may appear obvious, the temptation of sticking to the internals of the system when exporting the information is great. The experiences of the HELM project strongly suggest that the design phase of the DTDs requires a lot of time and a lot of thought; but the result is worth the time spent, since wrong decisions will greatly slow-down further developments.

Only time and further experiences can have the final word on our guidelines and solutions. Still we believe we have provided at least a comprehensive list of the common problems that must be faced.

References

1. M. Agrawal, N. Kayal, N. Saxena, "PRIMES in P", unpublished, August 2002, http://www.cse.iitk.ac.in/users/manindra/primality.ps.
2. A. Asperti, F. Guidi, L. Padovani, C. Sacerdoti Coen, I. Schena, "Mathematical Knowledge Management in HELM", in On-Line Proceedings of the First International Workshop on Mathematical Knowledge Management (MKM2001), RISC-Linz, Austria, September 2001.
3. A. Asperti, L. Padovani, C. Sacerdoti Coen, Schena, I., "HELM and the semantic Math-Web". Proceedings of the 14th International Conference on Theorem Proving in Higher Order Logics (TPHOLS 2001), 3-6 September 2001, Edinburgh, Scotland.
4. A. Asperti, B.Wegner, "MoWGLI - A New Approach for the Content Description in Digital Documents", in Proceedings of the Ninth International Conference on Electronic Resources and the Social Role of Libraries in the Future, Section 4, Volume 1.
5. O. Caprotti, H. Geuvers, and M. Oostdijk, "Certified and Portable Mathematical Documents from Formal Contexts", in On-Line Proceedings of the First International Workshop on Mathematical Knowledge Management (MKM2001), RISC-Linz, Austria, September 2001.
6. Y.Coscoy. "Explication textuelle de preuves pour le Calcul des Constructions Inductives", Phd. Thesis, Université de Nice-Sophia Antipolis, 2000.
7. A. Franke, M. Kohlhase, "MBase: Representing Knowledge and Context for the Integration of Mathematical Software Systems", to appear in Journal of Symbolic Computation.
8. J. Harrison, "Real Numbers in Real Applications", invited talk at Formalising Continuous Mathematics 2002 (FCM2002), 19th August 2002, Radisson Hotel, Hampton, VA, USA
9. M. Kohlhase, A. Franke, "MBase: Representing Knowledge and Context for the Integration of Mathematical Software Systems", Journal of Symbolic Computation 23:4 (2001), pp. 365 - 402.
10. Learning Technology Standards Committee of IEEE, "Draft Standard for Learning Object Metadata", July 15th 2002.
11. C. Muñoz, "Un calcul de substitutions pour la représentation de preuves partielles en théorie de types", PhD. Thesis, University Paris 7, 1997.
12. The OpenMath Esprit Consortium, "The OpenMath Standard", O. Caprotti, D. P. Carlisle, A. M. Cohen editors.

Managing Digital Mathematical Discourse

Jonathan Borwein[1] and Terry Stanway[2]

[1] CoLab, The Centre for Experimental and Constructive Mathematics,
Simon Fraser University, Burnaby, British Columbia, Canada. V5A 1S6
jborwein@cecm.sfu.ca,
http://www.cecm.sfu.ca/~jborwein
[2] CoLab, The Centre for Experimental and Constructive Mathematics,
Simon Fraser University, Burnaby, British Columbia, Canada. V5A 1S6
tstanway@cecm.sfu.ca

Abstract. In this paper, we propose that the present state of Mathematical Knowledge Management (MKM) derives from two main imperatives: the desire to organize and encapsulate mathematical knowledge after it is produced and the desire to encapsulate the act of production. While progress in digital network technology has facilitated a confluence of these efforts, their original separation imposes an approximate rubric on MKM which may be used to help define the challenges facing the field. We propose that one of the main challenges lies in the question of fixed versus flexible ontologies and the related question of ontology resolution between applications. Finally, we describe *Emkara*, an application which adopts a flexible metadata definition in the archiving and retrieval of digital mathematical exchanges, queries, and grey literature.

1 MKM's Intellectual Pedigree

Involving research mathematicians as well as specialists from such diverse fields as librarianship, education, cognitive science, and computer science, there are currently a wide range of initiatives and projects that may be considered as belonging to the field of Mathematical Knowledge Management (MKM). A perusal of the *Proceedings of the First International Workshop on Mathematical Knowledge Management* reveals that presentations with a focus on best practice in the exchange mathematical documents in digital environments, such as a presentation on the recommendations of the International Mathematics Union's *Committee on Electronic Information and Communication*, shared time with presentations focussed on foundational concepts, such as a presentation on the underlying logic and language of the *Theorema* theorem proving system.[1] The juxtaposition of topics represented by these two presentations represents a fundamental duality of focus in the field of MKM, as the intellectual foundations of the two presentations are distinct: those of the former stretching back to the

---

[1] The former presentation was by Dr. Jonathan Borwein and the latter by Dr. Bruno Buchberger. (Proceedings of the *First International Workshop on Mathematical Knowledge Management*)

A. Asperti, B. Buchberger, J.H. Davenport (Eds.): MKM 2003, LNCS 2594, pp. 45–55, 2003.

libraries of antiquity and those of the latter, while more recent, reaching back at least as far as Leibniz' seventeenth century call for a *calculus philosophicus*.[2]

From this perspective, the pre-history of MKM is the dual histories of mathematical librarianship and mathematical logic. While both of these are 'meta-fields' in the sense that both are *about* mathematics, it would not have been immediately obvious to a pre-digital intellect that they share anything else in common. That emerging computer and network related technologies have redefined these fields in such a way that there are now good reasons to consider them as aspects of a single field, is an example of how a shift in media can lead to a shift in perspective.[3] The benefit of analysing MKM's intellectual pedigree is that not only does it help bring into focus the field's central preoccupations but it also helps identify some underlying tensions. From the librarianship side, MKM has inherited a concern for preservation, metadata, cataloguing, and issues related to intellectual property and accessibility. From the mathematical logic side, MKM has inherited a concern for foundations and issues related to automated or guided proof generation. Both traditions have bequeathed a concern for authentication of knowledge, albeit in different contexts. One task that is a concern in both of these two founding fields but is treated differently in each is the question of how to establish the underlying semantics that any exercise in information sharing requires. From a knowledge management perspective, this is the question of *ontology definition* and in the following section, we examine some of the problems presented by ontology definition in MKM.

2   Ontology Definition

The philosophical concept of *domain ontology* has important implications for MKM. In their *Scientific American* article, *The Semantic Web*, Berners-Lee, Hendler, and Lassila define "ontology" in the context of artificial intelligence and web-based applications:

> ... *an ontology is a document or file that formally defines the relations among terms. The most typical kind of ontology for the Web has a taxonomy and a set of inference rules.*[4]

For our purposes, we will bear in mind this broad definition, but seek out a more precise description in order to address ontology problems in MKM. In particular, our focus will be on the *discourse* of mathematical communities as opposed to their literature and we will consider management problems arising

---

[2] *From Frege to Gödel: A Source Book in Mathematical Logic, 1879-1931*, Jean van Heijenoort (editor), (Boston: Harvard University Press, 1967) 6.

[3] The conference description for the *First International Workshop on Mathematical Knowledge Management* describes MKM as an "exciting new field in the intersection of mathematics and computer science".

[4] Berners-Lee, T. Hendler, J. Lassila, O. *The Semantic Web*, Scientific American, May 2001. 35-43.

from the informal exchange of information conducted in the shared vocabularies of communities that make up the broader mathematical community.

In *The Acquisition of Strategic Knowledge*, Thomas R. Gruber describes five overlapping stages of knowledge acquisition. These are: identification, conceptualization, formalization, implementation, and testing.[5] While originally conceived for the build-up of knowledge in expert systems, these five stages provide a useful framework for the description of MKM knowledge management tasks. In particular, the description of the conceptualization stage draws from the terminology of *ontological analysis*, specifying three distinct aspects of the broad ontology: static ontology, dynamic ontology, and epistemic ontology. Gruber describes this stage as follows:

> *Conceptualization results in descriptions of categories or classes of the domain objects and how objects are related (the static ontology), the operators, functions, and processes that operate on domain objects (the dynamic ontology), and how all this knowledge can be used to solve the application task (the epistemic ontology).*[6]

By way of an example, consider the application of the language of conceptualization stage ontological analysis to a typical MKM knowledge retrieval task: the discovery of publications which mention the *Bartle-Graves theorem* in their title, abstract, or keywords. In this case, the static ontology includes a definition of the 'publication', 'title', 'abstract', and 'keywords' entities as well as a definition of the entities to be searched. The dynamic ontology includes a definition of the protocols and processes involved in information access and retrieval; this could be as simple as a SQL search on a fixed database or a more complex specification such as that of an agent-based query of a remote database. The epistemic ontology defines the interface level entities, both style related and logic related which will be invoked to determine the manners in which the information request and results interfaces may be presented to the user. According to a defined set of criteria, if the knowledge acquisition cycle is effective, the user is presented with output that, in some useful manner, lists publications that are related to the *Bartle-Graves theorem* along with relevant background information regarding these publications. A more sophisticated epistemic ontology may present related information based on an inference concerning what type of information might be of use to a particular user.

The dual inheritance of MKM is reflected in the definition of static ontologies. Applications that draw more strongly from the librarianship tradition, admit degrees of flexibility and ambiguity in their ontologies. Both Math-Net's *MPRESS* and the NSF funded *arXiv* mathematical document servers admit weakly defined elements in their metadata sets, asking submitting authors to make subjective assignments of topic descriptors.[7] Applications that draw from mathematical

---

[5] Gruber, T. *The Acquisition of Strategic Knowledge*, Academic Press, 1989. 128
[6] Gruber, 128.
[7] In the case of MathNet, these descriptors are referred to as 'keywords' and in the case of arXiv, they are referred to as 'Mathematical Categories'.

logic depend on highly fixed ontologies. The static ontology of *Theorema* is encoded at the implementation level as highly structured 'Theorema Formal Text', an implementation of high order predicate logic.[8]

Negotiating differences in ontologies is part of human communication. It is therefore not surprising that applications that have evolved from the highly human-centred discipline of librarianship have inherited a tolerance for subjectivity and a degree of ambiguity. It is similarly not surprising that applications that have evolved out of the field of mathematical logic depend upon fixed and highly defined ontologies. Certainly, ontology resolution is a problem that must be addressed in any effort to interconnect MKM applications. While much work has been done in this area resulting in significant progress, notably, by the *Open-Math* and *OMDoc* research groups, much work remains. Consider, for example, the task of determining whether the proof of a given proposition either exists in the literature or can be automatically generated. Significant refinements of current technology are required before this determination can be reliably accomplished by a purely agent-based query of proof repositories or theorem proving systems. The task is made more difficult if the proposition is originally expressed using a human-centred application which accepts input, such as LaTeX or *Presentation MathML*, that maps directly to standard mathematical text. In this case, it is easy to imagine that some form of challenge and response interaction may be necessary in order to determine the semantic content of the query.

While the problems associated with ontology negotiation in MKM have received considerable attention, equally germane is the fundamental question of ontology construction. Motivated by the desire to build applications from a solid foundation in predicate logic, explicit attention to ontology construction has historically been a characteristic of research in artificial intelligence. This research has resulted in a number of languages and methods for building ontologies such as *Knowledge Interchange Format*.[9] Of note, in the domain of MKM, is the work done by Fürst, Leclère, and Trichet in developing a description of projective geometry based on the *Conceptual Graphs* model of knowledge representation.[10] Ontology definition is made more difficult if it is impossible to describe *a priori* aspects of the knowledge domain. This is precisely the case with *grey literature* in which elements of the static, dynamic, and epistemic ontologies will inevitably need to be extended as new fields and forms of knowledge are defined. In the next section, we examine the need for a flexible and extendable ontology in managing digital mathematical grey literature and mathematical discourse in general.

---

[8] Buchberger, Bruno. *Mathematical Knowledge Management in Theorema*, proceedings of *The First International Workshop on Mathematical Knowledge Management*

[9] http://logic.stanford.edu/kif/kif.html

[10] Fürst, Frédéric, Leclère, Michel, and Trichet, Francky. *Contribution of the Ontology Engineering to Mathematical Knowledge Management* proceedings of *The First International Workshop on Mathematical Knowledge Management*

3   The Digital Discourse

As mathematical activity is increasingly conducted with the support of digital network helper technologies, it is becoming increasingly possible to address the questions of to what extent, and for what purposes, this activity can be captured and archived. The types of mathematical activity that are conducted with the aid of digital networks cover a spectrum with highly informal activities such as queries to search engines and mathematical databases at one end and, at the other end, activities of a much more formal nature such as the publication of papers in online journals and preprint servers. As the majority of online mathematical activity is informal, the challenge that it presents to MKM is akin to the challenge presented by "grey literature" to the field of librarianship. A simple Web search reveals that the question of what constitutes grey literature is very much open to interpretation. Definitions range from the restrictive "theses and pre-prints only" to more inclusive interpretations such as the following from the field of medical librarianship:

> *In general, grey literature publications are non-conventional, fugitive, and sometimes ephemeral publications. They may include, but are not limited to the following types of materials: reports (pre-prints, preliminary progress and advanced reports, technical reports, statistical reports, memoranda, state-of-the art reports, market research reports, etc.), theses, conference proceedings, technical specifications and standards, noncommercial translations, bibliographies, technical and commercial documentation, and offcial documents not published commercially (primarily government reports and documents).[11]*

This idea of the "fugitive and sometimes ephemeral" nature of grey literature is particularly apt when applied to expression conveyed via digital networks. For purposes which will be discussed presently, we choose to adopt a definition of digital mathematical discourse (DMD) which encompasses the full *grey to white spectrum*. A list of examples from this spectrum might contain such diverse mathematical entities as email exchanges, bulletin boards, threaded discussions, CAS worksheets, transcripts of electronic whiteboard collaborations, exam questions, and database query strings as well as preprints and published papers. The motivation for this open-ended definition is the potential that methods of archiving and retrieving DMD hold for both gaining insight into the nature of mathematical activity and facilitating productivity in mathematical activity. We turn now to the problem of specifying an appropriate ontology development framework.

4   Ontology Development for Digital Mathematical Discourse

There are two reasons that careful consideration of an ontology development framework for DMD is important. The first relates to the "Web Services" aspect

---

[11] www.nyam.org/library/greylit/whatis.shtml

of the Semantic Web specification and the potential for DMD oriented applications to both harvest and be harvested from, either manually or via agents. The second is more complicated and is related to the emerging state of metadata standards. Jokela, Turpeinen, and Sulonen have argued that if an application's main functions are content accumulation and content delivery, then a highly structured ontology definition and corresponding logic is unnecessary. In this case, ontologies may effectively be defined by way of formally specified metadata structures.[12]

One problem confronting DMD ontology development is that while it would be enticing to simply make the ontology implicit in an *application profile* combining, for example, the *Dublin Core* and the related *Mathematics Metadata Markup* metadata specifications, it is not clear that such a profile would be rich enough or offer fine enough granularity to meet the needs of a DMD specification. A second problem originates from the objective that DMD applications be acessible to the broader mathematical community, including individuals with possibly limited understanding of the metadata standards at their disposal. A lack of understanding of metadata standards introduces the possibility of inappropriate use of taxonomy and the unnecessary use of ad hoc taxonomy. For these reasons, for the definition of new forms of mathematical expression, an appropriate ontology based on a profile of existing metadata standards, must exhaust those standards and then encourage the intelligent use of ad hoc taxonomy to complete the definition to the desired degree of granularity. As the dominant metadata standards become more fully refined, the ad hoc component should be reconciled with the existing metadata profile. *Figure 1* represents a component of the static ontology of the type of threaded discussion that might be motivated by an online class discussion; ad hoc elements are connected into the diagram by dotted lines. This ontology is based on an application profile that references the *Dublin Core* and *Educational Modeling Language* name spaces.

The extension of a metadata application profile by the addition of elements that are not defined in the schemata that make up the profile leads to potential management problems. These include incomplete object definition, in which not enough elements are defined to allow the desired degree of granularity, as well as contradictory or superfluous definition of elements. While it is certainly possible, depending on the size of the knowledge base, that these problems be resolved manually, it is also possible to at least partially automate the management process. In the next section, we will discuss some design issues for a DMD management application in the context of progress on the *Emkara* project at Simon Fraser University's *CoLab*.

---

[12] Jokela, Sami, Turpeinen, Marko, and Sulonen, Reijo, *Ontology Development for Flexible Content*, *Proceedings of the Hawaii International Conference on System Science 2*

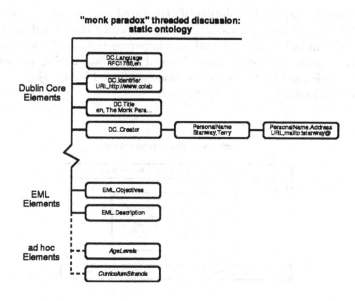

Fig. 1. *The Monk Paradox: Online Class Discussion*

5   The Emkara Project

The purpose of the Emkara project is to conduct investigations in DMD interfaces and management. A major component of the project is the design and construction of a DMD management application based on open source technology. Corresponding respectively to the static, dynamic, and epistemic ontologies, this application has three fundamental components: an archiving component which respects the static ontology, an object and interface building component that respects the dynamic ontology, and a user interfaces component which respects the epistemic ontology. Caste as design objectives, these three components reflect the intent to provide:

- qualified user control over granularity of object classification
- qualified user control over functionality and interface design, and ...
- end user access to knowledge creation and retrieval interfaces.

Secondary design objectives include the ability to store all forms of mathematical content and the ability to translate structured mathematical text between LaTeX, MathML, and, ultimately, OMDoc formats.

At the implementation level, an Emkara system consists of a front-end CSS and MathML compatible web interface together with a MathML editor configured as a "helper application". The mid-level implementation consists of a SQL compatible relational database management system, functions which support the creation and management of user interfaces, functions which support user authentication and session management, and functions which support data transformations such as translation of structured text formats. The back-end consists of the database tables and a data directory tree. Used primarily for the storage of metadata and internal system parameters, the tables also provide a convenient location to store data obtained from data mining processes. The directory tree is used for the storage of documents and large objects.[13] *Figure 2* illustrates the Emkara system architecture.

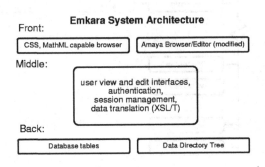

Fig. 2. *Emkara System Architecture*

While the prototype is intended mainly as an experimental interface, a fully implemented Emkara system presents opportunities for research which respond to each of the fundamental ontologies. As alluded to in the last section, making the static ontology open and subject to extension by qualified users presents design questions concerning how to present an ontology editing environment along with the question of how to ensure that the ontology is being edited in a useful manner. That the static ontology is being represented as XML makes testing for "well-formedness" a simple test for consistency however, such a test makes no comment about superfluousness or lack of detail. The latter failings may only become apparent with use. It is a valid question, therefore, whether or not data regarding user interaction with the system can be processed in such a way as to reveal strengths and weaknesses in the static ontology. If so, then it is possible that the process of ontology management can be at least partially automated.[14]

A compelling question related to the introduction of unneeded vocabulary in the construction of static ontologies concerns the potential differences between the language that members of a particular community of interest use to describe

---

[13] The prototype is implemented using Postgresql[TM] and PHP4[TM] with a lightly modified Amaya[TM] browser serving as the MathML editor helper application. The modified Amaya browser must be installed on the client along with a CSS and MathML compatible browser.

[14] We expect to profit from the considerable amount of work done by the members of the *Ontolingua* research team concerning ontology development environments and automated ontology analysis.

their field and the vocabulary presented by the relevant schemata. The latter are typically developed through extensive committee work and are designed to meet a much broader range of needs than those of an individual communicating ideas to fellow members of a given community. It is important that methods be developed to support the presentation of schemata vocabularies in such a way that users can identify the elements they need to express their ideas and make the appropriate association with the language of their particular communities.

The dynamic ontology underlies the relationship between entities in the static ontology and the way they interact with each other. A "flash card" object may consist of only two main fields: question and answer. The dynamic ontology allows for the description of how these two fields interact in the context of the flash card object and directly reflects the functions and methods that determine object behaviour.[15] A valid question concerns the numbers and types of general object interactions that are necessary in order to construct the type of behaviour required by DMD. A related question is that of ontology representation: how many and what types of object access methods must be exposed in order to afford qualified "non-programmers" the flexibility they need to create the behaviour they desire. With respect to these questions, inspiration, guidance, and, possibly, code, can be derived from the work of the members of the *Protege-2000* research team who have created a visual ontology editor and undertaken research into its application in the creation of ontologies in several Semantic Web compatible knowledge interchange languages.[16]

Questions that can be addressed at the level of the epistemic ontology include any question related to the type of interface appropriate for a particular task and user device. If the task is helping students understand the material in a particular lesson, a threaded discussion might be part of the solution. The related question is what access and acquisition methods make sense for different devices?

Related to the epistemic ontology, a number of valid research objectives are derived from the fact that an Emkara system is necessarily a *stateful* system which must enforce authentication and a system of maintaining read and edit access privileges. From this follow questions related to how users interact with the system and the system's function as a facilitator of "on-line community". In particular, users are instances of a "user object" that includes metadata that makes up a user profile. This permits the collection and processing of data regarding what types of users are accessing what types of data and using what kinds of interfaces.

One of the most compelling features of any system that attempts to provide management functions for digital mathematical discourse is the potential that the system has to interact with other MKM applications via the emerging Semantic Web interchange languages and Web Services protocols. While atomic in nature, search strings for pre-print servers and input to theorem provers are

---

[15] In the prototype, the dynamic ontology is bound up in PHP4 classes and scripts and therefore some knowledge of PHP is required to customize system behaviour.

[16] Noy, N.F. et al, *Creating Semantic Web Contents with Protege-2000*, IEEE Intelligent Systems, March-April 2001

forms of grey literature. If a DMD application is able to provide interfaces for remote access to these systems coupled with organized methods of storing and retrieving the output, it is possible that a significant proportion of access to MKM applications would take place via interfaces that could provide meaningful information about the ways in which mathematics is being accessed and shared through the Web. In this situation, the payoff would be data concerning what types of users are accessing what types of documents in pre-print servers and what types of users are using what types of services provided by theorem provers.

6   Conclusion and Future Work

Emkara is now situated in the *CoLab* at Simon Fraser University's *Centre for Experimental and Constructive Mathematics*.[17] Dedicated to the investigation of advanced digital collaboration in mathematics research and education, the *CoLab* provides the opportunity to experiment with different ways of representing and communicating mathematical thought with the aid of advanced network and visualization technologies. As *Emkara* based interfaces are further developed, they will provide a means of capturing and archiving this mathematical activity.

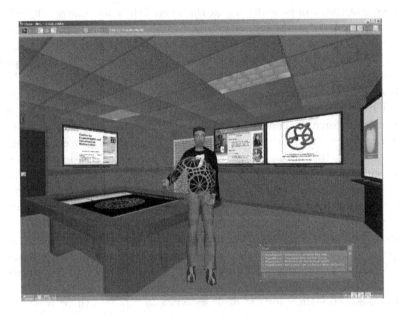

Fig. 3. *The (virtual) CoLab with avatar*

---

[17] Figure 3 is a virtual rendering of the CoLab by Steve Dugaro with the aid of Muse™ immersive 3D client software

Due largely to its diverse intellectual inheritance, the field of Mathematical Knowledge Management remains divided across a spectrum with applications whose main focus is derived from the field of mathematical librarianship at one end and applications whose main focus is derived from the field of mathematical logic at the other. In between, are applications such as Computer Algebra Systems, function libraries, and reverse look-up interfaces that interact with each other only with human intervention. By providing a means of interfacing with and organizing data from diverse MKM applications, digital mathematical discourse applications have the potential to exert a unifying influence on the field. Along with the challenges presented by ontology definition and interface design, future work with the *Emkara* system will focus on interaction with other MKM applications and on the interface design considerations associated with the support of online mathematical communities.

NAG Library Documentation

David Carlisle and Mike Dewar

NAG Ltd, Wilkinson house, Jordan Hill Road, Oxford, UK, OX2 8DR
{davidc,miked}@nag.co.uk

Abstract. This paper describes the management and evolution of a large collection of 1200 documents detailing the functionality in NAG Library products.

This provides a case study addressing many of the issues which concern the "MKM" project, involving conversion of legacy formats (SGML and LaTeX) to XML, and inferring semantic content from mainly presentational mathematical expressions.

1 The NAG Library Documentation

The NAG Fortran Library consists of around 1200 routines providing a range of mathematical functions. Each routine is accompanied by a document describing the mathematical problem being addressed and also the interface to the routine. PDF versions of these documents are freely available from the URL: http://www.nag.co.uk/numeric/fl/manual/html/FLlibrarymanual.asp

The documents have been written by a large number of different authors although fairly strict style and content guidelines have been enforced. The earliest documents possibly as much as twenty-five years old and were originally typed on a golfball typewriter before an electronic typesetting system (TSSD, a typesetting system for scientific documents developed by Harwell laboratories) was adopted. The documents were subsequently migrated to SGML and proprietary systems such as DynaText were used to deliver them to users until PDF became widespread at which point that was adopted as the main electronic delivery mechanism. One of the issues throughout this process has been how to mark-up and render mathematical objects and expressions: originally this was done via "ASCII art" (i.e. 2-D representations based on ASCII characters), then via TSSD's own native format, and finally by embedding LaTeX fragments inside the SGML.

In this paper we will discuss issues involved in migrating these documents from their most recent format (SGML with mathematics encoded as LaTeX fragments) into an XML-based format that allows the document sources to be more readily used for multiple purposes: generating documentation in other formats (HTML, XHTML+MathML), documenting other products (C library), generating interface code and header files. In principle we would like to adopt a totally content-based markup scheme but, given the amount of documentation involved and the critical requirement of maintaining the correctness of the documents, the migration from presentation to content must be automated. This is much harder than it might first appear and we will describe some of the problems encountered (and the solutions adopted) in what follows.

A. Asperti, B. Buchberger, J.H. Davenport (Eds.): MKM 2003, LNCS 2594, pp. 56–65, 2003.

2   Mathematical Fragments in SGML Documents

The NAG documentation naturally includes many mathematical expressions. At the conclusion of this project the number of mathematical expressions is 112,830 (many of which appear in XML entities and are re-used in a number of different documents). Before this project, for the reasons explained below, this number was substantially higher.

Mathematical expressions were previously encoded using a single element. The content of this element was declared as #PCDATA, that is, it was essentially a string. The content of the string was a LaTeX expression. Thus a typical fragment might look like

```
<p>.... <maths>\frac{x}{2}</maths> ....
```

Many SGML systems have a native ability to handle such TeX fragments (Dyna-Text, 3B2, ...). Moreover, if the documents are to be printed via conversion to LaTeX, then any such fragments may be directly copied into the TeX document resulting from translating the SGML markup. (More details on the TeX and LaTeX typesetting systems may be obtained from the TeX Users group [3].)

While such a dual markup solution is apparently easy to define and can produce high-quality printed documentation, the lack of integration of the mathematical structure with the markup structure of the non-mathematical text causes problems when producing more structured output, for example on-line hypertext documentation in PDF or HTML. The NAG documentation will frequently make reference to a parameter (argument) of a function/subroutine. The SGML markup for a parameter A is `<arg>A</arg>`. In printed documentation this just produces a roman 'A' but in online text the 'A' is an active link back to the main documentation of the parameter. However if mathematical fragments are restricted to #PCDATA then the `<arg>A</arg>` markup may not be used. Instead one uses an equivalent TeX expression such as

```
<math>\mathrm{A} + \mathrm{B} > 0</math>
```

or artificially breaks up the mathematical expression to allow the SGML markup to be used:

```
<arg>A</arg><math>{}+{}</math><arg>B</arg><math>{} > 0</math>
```

The first method provides acceptable printed output but hyperlinking is lost, also the creation of, for example, automatic indices of use of parameters is made harder as the markup of function arguments is not consistent. The second method is only feasible if the arguments appear at the top level of the expression, not as a parameter of a TeX macro such as `\frac{...}{...}` or `\sqrt{...}`. It is also hard to ensure that the resulting expression retains the correct spacing (note the empty brace groups in the above) and most importantly it obscures the mathematical structure of the expression (which is now encoded as two separate mathematical fragments). This makes it harder to automatically infer any meaning from the markup, or convert to other formats such as Content MathML [1] or OpenMath [2] which would be able to encode the sum and inequality in semantically richer formats.

Even if we could overcome these problems and amalgamate all such mathematical fragments into complete objects, there will still be many cases where the mathematical objects have no obvious semantic meaning or cannot be represented in a content language. For example the expression:

```
<maths><arg>MAXIP</arg>\geq 1\ +</maths>
    the number of values of <maths><arg>ISX</arg>&gt;0</maths>
```

is perfectly reasonable but on their own the two mathematical expressions are meaningless.

NAG has taken a gradual approach to improving this situation. In the first stage (which was achieved for the Mark 20 Fortran library) the SGML sources were converted to XML, and the XML Document Type Definition (DTD) was extended to allow certain elements within mathematics. So the above could be encoded more naturally:

```
<math><arg>A</arg> + <arg>B</arg> > 0</math>
```

The markup is also allowed inside TEX expressions such as

```
<math>\frac{<arg>A</arg>}{2}</math>
```

This intertwining of TEX and XML is also not ideal and complicates some of the further processing of the documents. An eventual aim is to convert to MathML, with additional elements (such as `<arg>`) from the NAG DTD. In this case the fraction would be encoded as

```
<math><mfrac><arg>A</arg><mn>2</mn></mfrac></math>
```

and the inequality above would be encoded as

```
<math><mrow><arg>A</arg> <mo>+</mo> <arg>B</arg></mrow>
                    <mo>&gt;</mo>   <mn>0</mn></math>
```

Having the structure in MathML (even Presentation MathML rather than the semantically richer Content MathML markup) would greatly ease further processing, however currently the documents are primarily hand authored and it is unreasonable to hand author the more verbose MathML markup. Experiments are continuing with available MathML editing tools and also with automatic conversion from TEX-like syntax.

3 Classifying Mathematical Expressions

In printed documentation the expression "A(1)" might denote either the first element of an array A or the function A applied to the number 1. As XML markup may be used within mathematical expressions these may be distinguished. Array references are marked using NAG-specific elements, that markup both the array reference (`<ar>`) and the array index(`<ai>`). Markup of the form

```
<ar><arg>A</arg><ai>1</ai></ar>
```

will be typeset as A(1) in the documentation of the Fortran library, with the A being a link to the description of the array valued parameter "A" if the documentation format supports links. The improved structure of the documentation now allows documentation of the C interface to the routine to be derived from the same source. The C library documentation would typeset the same markup as a[0] as C array indexing is 0-based, and the C library conventionally uses lower case names. The recent Mark 7 release of the NAG C library contains around 400 new functions. The documentation of each of these is automatically generated from the same source as the documentation of the corresponding Fortran routine. In most cases the code of the C interface is also generated from the same file.

Moving from essentially unstructured TEX markup to this structured form that allows arrays to be identified and re-indexed has been a major effort and cannot be fully automated. A great deal of in-house knowledge was used in the conversion (for example A(...) in the linear algebra chapters F07 and F08 *always* refers to a matrix not a function, while in the statistics chapters P(...) usually denotes a probability function). The generated C library documentation went through many cycles of proof reading, at each stage any errors reported were used to further refine the conversion scripts. This iterative procedure appears to be successfully producing highly structured and accurate documentation, however the tools have been highly customised for this one documentation set. The more general problem, highlighted in the MKM project description of converting "legacy" TEX documents may require quite different techniques if one does not have extra knowledge about the subject area of the document.

3.1   Constraints

Mathematical expressions occur primarily in two places in the library documentation. Firstly in the description section of each routine, where the mathematical functionality is described. Secondly in the description of each parameter where many constraints on the parameter values are documented. It has proved especially useful to have these constraints in a strongly marked up form. Having fully marked up the constraints it has been possible to automatically generate C code for the C library that checks the constraints and generates appropriate error messages if the user input violates these constraints. Thus a constraint appearing in the XML as

```
<con><maths><arg>M</arg> \geq 0</maths></con>
```

will appear in the Fortran documentation as

$$M \geq 0$$

in the C documentation as

$$m \geq 0$$

and in the C library sources as

```
  if (!(m >= 0)) {
  ....
  v=sprintf(buf, nag_errlist[NE_INT], "m", m, "m >= 0");
```

4   Classifying Function Arguments

In addition to using more structured markup within mathematical expressions, the main other enhancement to the library XML which has allowed the documents to be used for multiple products is the additional markup used to classify the parameters of the routine. The aim has been to get away from the idea that the document is a human-readable description of how to use the software, and instead treat it as a high-level specification of the algorithm and its parameters.

The description of each parameter is prefixed by an element whose attributes classify the parameter. For example the parameter M used in the constraint above (in the routine F08AEF) is classified by

```
<paramhead xid="M"
    type="integer" class="I" purpose="data"/>
```

That is it is an input (I) parameter of integer type which forms part of the user's problem description. This is used to generate the appropriate Fortran and C variable declarations

In the same routine, the Fortran interface makes use of a "workspace array"

```
<paramhead xid="WORK"
    type="rarray" class="0" dim1="*" purpose="workspace"/>
```

i.e. a real valued array which must be allocated by the program that calls this subroutine, and which must be at least a certain size, which will be documented in a constraint in the parameter description. In the C Library such workspace parameters do not appear in the function interface. Code is automatically generated to allocate the required space.

Almost all the parameters in the C interface may be similarly automatically constructed from the algorithm specification, once sufficient 'class' and 'purpose" attributes have been added. There are a few cases where this automatic conversion needs to be overridden by explicitly supplying the description of a parameter in the C interface, but this mechanism proved to be very rarely needed, most often in cases where the C interface needed to be kept consistent with earlier releases that were not so closely modeled on the Fortran interface.

The resulting documents may be viewed on the NAG website
Fortran: http://www.nag.co.uk/numeric/fl/manual/pdf/F08/f08aef.pdf
C: http://www.nag.co.uk/numeric/cl/manual/pdf/F08/f08aec.pdf
The first two pages of each document are shown in Figure 1 and 2.

5   Conclusions

What we have described is a real-life case study of the transformation of an existing body of legacy documentation to make its inherent structure and semantics explicit. The project was driven by commercial imperatives and successfully met its goals.

F08 – Least-squares and Eigenvalue Problems (LAPACK) **F08AEF (SGEQRF/DGEQRF)**

NAG Fortran Library Routine Document

F08AEF (SGEQRF/DGEQRF)

Note: before using this routine, please read the Users' Note for your implementation to check the interpretation of ***bold italicised*** terms and other implementation-dependent details.

1 Purpose

F08AEF (SGEQRF/DGEQRF) computes the QR factorization of a real m by n matrix.

2 Specification

```
SUBROUTINE F08AEF(M, N, A, LDA, TAU, WORK, LWORK, INFO)
ENTRY      sgeqrf (M, N, A, LDA, TAU, WORK, LWORK, INFO)
INTEGER           M, N, LDA, LWORK, INFO
real              A(LDA,*), TAU(*), WORK(*)
```

The ENTRY statement enables the routine to be called by its LAPACK name.

3 Description

This routine forms the QR factorization of an arbitrary rectangular real m by n matrix. No pivoting is performed.

If $m \geq n$, the factorization is given by:

$$A = Q\begin{pmatrix} R \\ 0 \end{pmatrix},$$

where R is an n by n upper triangular matrix and Q is an m by m orthogonal matrix. It is sometimes more convenient to write the factorization as

$$A = \begin{pmatrix} Q_1 & Q_2 \end{pmatrix}\begin{pmatrix} R \\ 0 \end{pmatrix},$$

which reduces to

$$A = Q_1 R,$$

where Q_1 consists of the first n columns of Q, and Q_2 the remaining $m - n$ columns.

If $m < n$, R is trapezoidal, and the factorization can be written

$$A = Q\begin{pmatrix} R_1 & R_2 \end{pmatrix},$$

where R_1 is upper triangular and R_2 is rectangular.

The matrix Q is not formed explicitly but is represented as a product of $\min(m,n)$ elementary reflectors (see the F08 Chapter Introduction for details). Routines are provided to work with Q in this representation (see Section 8).

Note also that for any $k < n$, the information returned in the first k columns of the array A represents a QR factorization of the first k columns of the original matrix A.

4 References

Golub G H and van Loan C F (1996) *Matrix Computations* (3rd Edition) Johns Hopkins University Press, Baltimore

[NP3546/20A] *F08AEF (SGEQRF/DGEQRF).1*

Fig. 1. Fortran interface

5   Parameters

1:     M – INTEGER   *Input*

On entry: m, the number of rows of the matrix A.

Constraint: $M \geq 0$.

2:     N – INTEGER   *Input*

On entry: n, the number of columns of the matrix A.

Constraint: $N \geq 0$.

3:     A(LDA,*) – *real* array   *Input/Output*

Note: the second dimension of the array A must be at least $\max(1, N)$.

On entry: the m by n matrix A.

On exit: if $m \geq n$, the elements below the diagonal are overwritten by details of the orthogonal matrix Q and the upper triangle is overwritten by the corresponding elements of the n by n upper triangular matrix R.

If $m < n$, the strictly lower triangular part is overwritten by details of the orthogonal matrix Q and the remaining elements are overwritten by the corresponding elements of the m by n upper trapezoidal matrix R.

4:     LDA – INTEGER   *Input*

On entry: the first dimension of the array A as declared in the (sub)program from which F08AEF (SGEQRF/DGEQRF) is called.

Constraint: $LDA \geq \max(1, M)$.

5:     TAU(*) – *real* array   *Output*

Note: the dimension of the array TAU must be at least $\max(1, \min(M, N))$.

On exit: further details of the orthogonal matrix Q.

6:     WORK(*) – *real* array   *Workspace*

Note: the dimension of the array WORK must be at least $\max(1, LWORK)$.

On exit: if INFO = 0, WORK(1) contains the minimum value of LWORK required for optimum performance.

7:     LWORK – INTEGER   *Input*

On entry: the dimension of the array WORK as declared in the subprogram from which F08AEF (SGEQRF/DGEQRF) is called, unless LWORK = −1, in which case a workspace query is assumed and the routine only calculates the optimal dimension of WORK (using the formula given below).

Suggested value: for optimum performance LWORK should be at least $N \times nb$, where nb is the **blocksize**.

Constraint: $LWORK \geq \max(1, N)$ or LWORK = −1.

8:     INFO – INTEGER   *Output*

On exit: INFO = 0 unless the routine detects an error (see Section 6).

Fortran interface (cont)

f08 – Least-squares and Eigenvalue Problems (LAPACK)   **f08aec**

NAG C Library Function Document

nag_dgeqrf (f08aec)

1   Purpose

nag_dgeqrf (f08aec) computes the QR factorization of a real m by n matrix.

2   Specification

```
void nag_dgeqrf (Nag_OrderType order, Integer m, Integer n, double a[],
    Integer pda, double tau[], NagError *fail)
```

3   Description

nag_dgeqrf (f08aec) forms the QR factorization of an arbitrary rectangular real m by n matrix. No pivoting is performed.

If $m \geq n$, the factorization is given by:

$$A = Q \begin{pmatrix} R \\ 0 \end{pmatrix},$$

where R is an n by n upper triangular matrix and Q is an m by m orthogonal matrix. It is sometimes more convenient to write the factorization as

$$A = (Q_1 \quad Q_2) \begin{pmatrix} R \\ 0 \end{pmatrix},$$

which reduces to

$$A = Q_1 R,$$

where Q_1 consists of the first n columns of Q, and Q_2 the remaining $m - n$ columns.

If $m < n$, R is trapezoidal, and the factorization can be written

$$A = Q(R_1 \quad R_2),$$

where R_1 is upper triangular and R_2 is rectangular.

The matrix Q is not formed explicitly but is represented as a product of $\min(m, n)$ elementary reflectors (see the f08 Chapter Introduction for details). Functions are provided to work with Q in this representation (see Section 8).

Note also that for any $k < n$, the information returned in the first k columns of the array **a** represents a QR factorization of the first k columns of the original matrix A.

4   References

Golub G H and Van Loan C F (1996) *Matrix Computations* (3rd Edition) Johns Hopkins University Press, Baltimore

5   Parameters

1: **order** – Nag_OrderType   *Input*

   On entry: the **order** parameter specifies the two-dimensional storage scheme being used, i.e., row-major ordering or column-major ordering. C language defined storage is specified by **order = Nag_RowMajor**. See Section 2.2.1.4 of the Essential Introduction for a more detailed explanation of the use of this parameter.

   Constraint: **order = Nag_RowMajor** or **Nag_ColMajor**.

Fig. 2. C interface

2:     **m** – Integer   *Input*

On entry: m, the number of rows of the matrix A.

Constraint: **m** ≥ 0.

3:     **n** – Integer   *Input*

On entry: n, the number of columns of the matrix A.

Constraint: **n** ≥ 0.

4:     **a**[dim] – double   *Input/Output*

Note: the dimension, dim, of the array **a** must be at least max(1, **pda** $\times$ **n**) when **order** = **Nag_ColMajor** and at least max(1, **pda** $\times$ **m**) when **order** = **Nag_RowMajor**.

If **order** = **Nag_ColMajor**, the (i, j)th element of the matrix A is stored in **a**[$(j-1) \times$ **pda** $+ i - 1$] and if **order** = **Nag_RowMajor**, the (i, j)th element of the matrix A is stored in **a**[$(i-1) \times$ **pda** $+ j - 1$].

On entry: the m by n matrix A.

On exit: if $m \geq n$, the elements below the diagonal are overwritten by details of the orthogonal matrix Q and the upper triangle is overwritten by the corresponding elements of the n by n upper triangular matrix R.

If $m < n$, the strictly lower triangular part is overwritten by details of the orthogonal matrix Q and the remaining elements are overwritten by the corresponding elements of the m by n upper trapezoidal matrix R.

5:     **pda** – Integer   *Input*

On entry: the stride separating matrix row or column elements (depending on the value of **order**) in the array **a**.

Constraints:

  if **order** = **Nag_ColMajor**, **pda** $\geq$ max(1, **m**);
  if **order** = **Nag_RowMajor**, **pda** $\geq$ max(1, **n**).

6:     **tau**[dim] – double   *Output*

Note: the dimension, dim, of the array **tau** must be at least max(1, min(**m**, **n**)).

On exit: further details of the orthogonal matrix Q.

7:     **fail** – NagError *   *Output*

The NAG error parameter (see the Essential Introduction).

6   Error Indicators and Warnings

NE_INT

  On entry, **m** = $\langle value \rangle$.
  Constraint: **m** ≥ 0.

  On entry, **n** = $\langle value \rangle$.
  Constraint: **n** ≥ 0.

  On entry, **pda** = $\langle value \rangle$.
  Constraint: **pda** > 0.

NE_INT_2

  On entry, **pda** = $\langle value \rangle$, **m** = $\langle value \rangle$.
  Constraint: **pda** $\geq$ max(1, **m**).

C interface (cont)

The main conclusion that we draw from this project is that it *is* possible to take a large collection of documents and infer a useful amount of content from the presentation markup *but* that this requires a large amount of specific information about the documents themselves. The benefits however, which include hyperlinking, more intelligent indexing and searching, better rendering in different target environments and the potential for re-use in un-anticipated contexts (in our case generating C library code and documentation from documents originally intended to document a Fortran library), may well make such an exercise worthwhile.

In our case we adopted an iterative approach rather than trying to produce tools which would do everything in one go. By making the conversion in discrete steps our documentation collection was at all times kept valid to a DTD (which changed at each step) and we were able gradually to adapt our in-house tools to use the new features as they were added. Thus while the general problem of TeX to Content MathML or OpenMath is probably unsolvable, and the problem of TeX to Presentation MathML is in general still hard, by using extra knowledge of the subject matter in the documents, and by allowing existing TeX based markup to remain in intermediate versions, extra semantic information has been added while keeping the document source in a form that may be processed at all times.

The success of our project depended on both the explicit structure of the existing documents (basically the SGML DTD) and the implicit structure (for example we know that in general every workspace parameter should have a minimum size described in the documentation). The same should be true of other kinds of mathematical documents, although the implicit structures which are important will of course vary. Any tools for migrating legacy documents to more semantically-rich forms will need to be highly customisable and support a great deal of human interaction.

Further details on the format used and possible future uses of the documents may be found in our paper [4]

References

1. D. Carlisle, P. Ion, R. Miner, N. Poppelier; (editors) Mathematical Markup Language (MathML) 2.0 Specification. http://www.w3.org/TR/MathML2
2. O. Caprotti, D. P. Carlisle, A. M. Cohen (editors); The OpenMath Standard, February 2000. http://www.openmath.org/standard
3. The TeX User's Group. http://www.tug.org
4. D. Carlisle and M. Dewar, From Mathematical servers to mathematical Services, Proceedings of the First International Congress of Mathematical, Editors, Arjeh Cohen, Xiao-Shan Gao, Nabuki Takayama, World Scientific, 2002, ISBN 981-238-048-5 Software,

On the Roles of LaTeX and MathML in Encoding and Processing Mathematical Expressions

Luca Padovani

Department of Computer Science
Mura Anteo Zamboni 7, 40127 Bologna
Italy
lpadovan@cs.unibo.it

Abstract. The Mathematical Markup Language (MathML [1]), a standard language for the encoding of mathematical expressions, is going to have a deep impact on a vast community of users, from people interested in publishing scientific documents to researchers who seek new forms of communication and management of mathematical information. In this paper we survey the worlds of LaTeX [10,11], a well-established language for typesetting, and MathML, whose diffusion is promisingly taking off in these days, with an emphasis on the management of mathematics. We will try to understand how they relate to each other, and why we will still need both in the years to come.

1 Introduction

The history of LaTeX can be traced back to 1977, when D. E. Knuth started designing METAFONT [4] and TeX [6,7,3]. METAFONT is a language for the mathematical description of font glyphs, while TeX is a language for arranging glyphs together, so that they form words, sentences, and mathematical formulae. In the mid-1980s, L. Lamport developed LaTeX, a set of macros that allowed to typeset with TeX at a higher level of abstraction, making the system easier to use and ultimately successful. In 2001 the Math Working Group published the specification of the Mathematical Markup Language Version 2.0 [1], an XML language whose aim is to encode mathematical expressions capturing both their appearance (presentation markup) and their meaning (content markup).

Just looking at the implementation page for MathML[1] we realize that the terms MathML, TeX, and LaTeX are almost equally distributed. Some of the tools presented are stylesheets, others are scripts, others are real, complex programs, many of them converting from LaTeX to MathML, a few doing the opposite. People started doing research on this subject, investigating how TeX and its macro mechanism relate to MathML [9]. At the same time, other people still have confused ideas about MathML and are somehow reluctant to "throw away" their knowledge of LaTeX just to adopt a young technology which has not stopped evolving yet.

---

[1] See http://www.w3.org/Math/implementations.html.

A. Asperti, B. Buchberger, J.H. Davenport (Eds.): MKM 2003, LNCS 2594, pp. 66–79, 2003.

So what is the direction of these developments? Is there a role for both LATEX and MathML in the Web era? What can we expect from MathML that we have not achieved with LATEX after so many years? Admittedly, LATEX and MathML do not lay on the same level, for the former allows typesetting of whole documents, whereas MathML is concerned with mathematical expressions only. We will restrict our analysis to the LATEX markup for expressions, being conscious that the challenging complexity of encoding mathematics can be considered a good basis to compare different technologies in general. We are not trying to follow a scheme or to draw an exhaustive comparison. Rather, we aim at giving the reader a glimpse of the differences and the similarities of the two languages by focusing on a few examples, hoping to clarify some points and to stimulate curiosity and motivations for further research.

The starting point of our discussion is the following mathematical formula, which will accompany us, in one form or another, for the rest of the paper:

$$\epsilon_0 \oint_S \mathbf{E} \cdot d\mathbf{A} = q \ . \tag{1}$$

Equation 1 is the Gauss's law as presented in D. Halliday, *et al.* [8]. In a different form it is better known as the first Maxwell equation, but we stick with this less popular version since it will prove more interesting to encode. In a nutshell, the equation states that the total of the electric flux out of a closed surface S multiplied by the the electric permittivity in the region of the electric field is equal to the charge enclosed within the same surface. The region is assumed to be a vacuum. In the formula $\oint$ denotes a surface integral.

The discussion that follows will concentrate on *syntax, semantics, formatting quality, macros and transformations.*

2   Syntax

The syntax of TEX and LATEX was born with a precise idea in mind: to make it easy for a human to type the markup, and to provide access to a wide variety of mathematical symbols solely by means of characters available on common computer keyboards. As D. E. Knuth reports in [5], "experience shows that untrained personnel can learn how to type [TEX] without difficulty" in a short amount of time.

The "pretty-printed" LATEX markup used to typeset equation 1 is shown in Fig. 1. The fact that the markup is so compact and so simple to type, the outcome still being such a beautifully formatted formula, is because LATEX takes care of most of the "dirty jobs:" space is automatically added around operators, the details about font size reduction and vertical displacement of scripts are coded in the formatting semantics of the scripting operator _. Also, the Computer Modern fonts that LATEX typically comes with have been explicitly designed to be used in conjunction with TEX, and a deep knowledge of the fonts used by a system for typesetting is a fundamental requirement for achieving professional results [16].

```
\begin{displaymath}
  \epsilon_0
  \oint_S
    \mathbf{E}
    \cdot
    d\mathbf{A} = q
\end{displaymath}
```

Fig. 1. LaTeX markup for equation 1. The `displaymath` environment surrounds a formula to be typeset in "display mode", in a paragraph of its own. Symbols and letters that are not normally available on keyboards are typeset using macros like `\epsilon` and `\cdot`, and bold letters are typeset using the macro `\mathbf`. The underscore symbol has the special meaning "subscript", so that the immediately subsequent symbol, or subformula, is formatted in subscript position. Braces { and } group subexpressions

```
<math xmlns="http://www.w3.org/1998/Math/MathML"
      display="block">
  <mrow>
    <msub>
      <mi> &epsi; </mi>
      <mn> 0 </mn>
    </msub>
    <mo> &InvisibleTimes; </mo>
    <mrow>
      <mrow>
        <msub>
          <mo xref="i1"> &oint; </mo>
          <mi> S </mi>
        </msub>
        <mrow>
          <mi> &#x1D404; </mi>
          <mo> &sdot; </mo>
          <mrow>
            <mo> &DifferentialD; </mo>
            <mi> &#x1D400; </mi>
          </mrow>
        </mrow>
      </mrow>
      <mo> = </mo>
      <mi> q </mi>
    </mrow>
  </mrow>
</math>
```

Fig. 2. MathML presentation markup encoding equation 1. XML documents are made of markup and Unicode characters [15,2]: the names `epsi`, `InvisibleTimes`, `oint`, `sdot`, `DifferentialD` are just human readable references (*entity references*, in XML jargon) for the Unicode characters U+03B5, U+2062, U+222E, U+22C5, U+2146. A character reference such as `&#x1D404;` refers directly to the Unicode character U+1D404

MathML is an XML application. As such it inherits the syntax of XML documents, which is equally well-known today due to the diffusion of markup languages for the Web, and XML itself. Since the very beginning of its specification, MathML does not even try to deceive its potential users: "while MathML is human-readable, it is anticipated that, in all but the simplest cases, authors will use equation editors, conversion programs, and other specialized software tools to generate MathML."

The first impression looking at the markup in Fig. 2, which encodes equation 1 in MathML presentation, is that of a dramatic difference with respect to LaTeX markup. A closer look reveals that things are not *that* different. The emphasis of MathML is on the *structure* of the formula. Not only the whole fragment is enclosed inside a `math` element, but every subexpression (`mrow`), including scripts (`msub`), is explicitly delimited by MathML tags. The atomic entities (called *tokens* in MathML) such as numbers, identifiers, operators are all marked up with corresponding elements `mn`, `mi`, `mo`. By its own nature, XML calls for embedding of different markup languages within each other. The `xmlns` attribute in the `math` element identifies the whole markup enclosed in `math` as MathML markup, and its value is the MathML *namespace*. XML namespaces allow any processor of XML markup to identify the parts of a document to be processed, and the suitable software components needed to process them.

If we abstract the syntax, the two encodings are very close and the layout schemata of the two languages are basically the same, as Table 1 summarizes. But the concrete representation of the structure and the explicit representation of the atomic components that in TeX and LaTeX are only implicitly (or heuristically) assumed are two strong points favoring MathML. The Web, where resource exploitation largely depends on the addressability of the resource itself, is made of links, and XML markup enables linking capabilities much more expressive and fine-grained than those available with LaTeX: we moved from the possibility of linking "documents" to the possibility of linking any structured fragment of any document.

3   Semantics

TeX and LaTeX are purely presentational languages whereas MathML is concerned with both semantics and presentation. We will come to MathML content markup in a moment, but first we have to fully appreciate presentation markup because, despite of the name, it conveys lots of semantics.

3.1   Semantics in Presentation

Structure. Structure is semantics. The multiplication sign · is an operator inside an `mrow` which is *more deeply nested* than the `mrow` that contains the equal sign =. Hence, we know that that the first operator (multiplication) has greater precedence with respect to the second operator (equality) *before* we know the actual names of the two operators. This is fundamental because processors of

Table 1. Correspondence between LaTeX and MathML layout schemata

Layout Schemata	LaTeX	MathML presentation
Identifiers	a,\sin,...	mi
Numbers	0,...	mn
Operators	+,(,\oint,...	mo
Grouping	{ }	mrow
Fractions	\frac	mfrac
Radicals	\sqrt	msqrt, mroot
Scripts	_, ^	msub, msup, msubsup, mmultiscripts
Stacked expressions, lines and braces above and below formulae	\stackrel, \underline, \overline, \underbrace, \overbrace	munder, mover, munderover
Matrices, tables	\begin{array}	mtable

MathML markup do not necessarily know mathematics as we do. Actually, we would like them to know as little about mathematics as possible, because the more rules (assumptions, conventions, heuristics) we hard-code in them, the more likely it is to find examples where the processors behave badly.

Let us stress the point with an example by considering a formatting engine for MathML that is capable of automatic line-breaking of very long formulae. Breakpoints are typically placed just before or after operators. The formatting engine will tend to favor breakpoints around operators that are "on the surface", rather than breakpoints around operators that are deeply nested. In this algorithm the formatting engine may ignore completely the actual names of those operators. By contrast, in the very few circumstances where TeX allows a line-break in a formula, the decision is solely based on the "name" of the operator, as any attempt of grouping subformulae prevents line-breaking. In fact an expression like $(1 + 1 + 1 + \cdots + 1)^2$ has a chance of being broken just because it is marked up as (1+1+1+\cdots+1)^2, where the superscript 2 applies to the closed parenthesis alone, and not to the whole subformula!

Unicode characters. Text inside a MathML document is made of Unicode characters, and Unicode classifies characters according to their semantics, not to their shape.[2] Let us compare **E** and **A** of the Gauss's law in the two encodings. In LaTeX they are encoded as \mathbf{E} and \mathbf{A}, where \mathbf's purpose is to set the bold font for its argument. In MathML they are encoded as the Unicode characters U+1D404 and U+1D400, which are defined as 'mathemat-

---

[2] Because of the unbound variety of meanings that mathematical symbols may have, Unicode had to make an unavoidable compromise on them [2].

ical bold capital A' and 'mathematical bold capital E'. Thus, the first encoding does not distinguish style from semantics, whereas the second encoding reminds us that **E** and **A** are precise mathematical objects, vectors in this case. MathML allows the application "styles," but with a different mechanism that does not generate such kind of ambiguities. A similar thing happens with the d that stands for the differential operator. In LaTeX it is just a plain letter d, whereas in MathML it has been encoded as the `DifferentialD` entity which, in turn, expands to the Unicode character U+2146, classified by Unicode as a *letter-like symbol*. It is a symbol, even if it resembles a letter.

Incidentally, if we look up the glyph that the book editor used for the surface integral we discover that in Unicode it corresponds to U+222E, 'contour integral,' and not to the surface integral operator which is U+222F and which has a different glyph. This is an interesting case where the publisher has chosen a *notation* that violates somehow the semantics of the formula. There can be many reasons for doing so, for example the search of consistency with other related documents that are typeset using that glyph, and not the right one. Either way, a document typeset using LaTeX usually requires the reader to give the "right" semantics to the symbols he sees by understanding the context. This is a problem though if the document is meant to be processed automatically, or if the formula is extracted from its context and communicated elsewhere. We delay further discussion of this point until Sect. 3.2, where we show how MathML content markup can be used to "restore" the semantics of a formula, whenever presentation is not enough.

No shortcuts. In mathematics the multiplication sign is omitted in most contexts. In equation 1 this convention has been applied between ϵ_0 and the surface integral because the multiplication involves two numbers, but not between **E** and d**A**, where the dot indicates multiplication of vectors. No matter what the reasons are to use one notation instead of another, in TeX and LaTeX the invisible multiplication is represented with... no markup!

In MathML presentation markup a multiplication operator is always used, even when the multiplication sign is invisible (the Unicode character U+2062 has an empty glyph). Other invisible operators that must be explicitly inserted in MathML are "apply function" (U+2061), separating a function from its argument as in $\sin x$ or $\sin(x)$, and "invisible comma" (U+200B), separating multiple indices, as in a script x_{ij}.[3] By allowing such verbose encoding of formulae, presentation markup conveys more meaning and (perhaps surprisingly) its rendering can be more precise. Let us consider "invisible times:" Computer Modern glyphs for letters in mathematical mode have been designed so that when they stand next to each other they look perfectly fine, as if they were multiplied together, but this might not be the case in other font families. If the markup has an explicit multiplication operator, as in MathML, the rendering engine can adjust the space between the two factors automatically. If not, it is up to the typist to be careful

---

[3] There are arguments favoring the definition of an "invisible plus" character, to be used in expressions like $2\frac{1}{4}$ when they mean $2 + 1/4$.

about the fonts he uses. The "apply function" operator is even more interesting: when we typeset expressions like $\sin x$ or $\sin(x)$ in LATEX we do not mind putting a space between the function and its argument if the argument has no fences around. LATEX takes care of this automatically, because it "knows" the function $\sin$. But a rendering engine for MathML can do it more generally, whenever the "apply function" operator occurs.

In the discussion above we have made the implicit assumption that rendering means *visual* rendering. But the use of explicit operators even when they have no visual appearance enables MathML with *aural* rendering as well, in a very natural way.

```
<math xmlns="http://www.w3.org/1998/Math/MathML">
  <apply>
    <eq/>
    <apply>
      <times/>
      <apply>
      <csymbol definitionURL="..."> ezero </csymbol>
      <apply>
        <int id="i1" definitionURL="..."/>
        <bvar>
          <ci> A </ci>
        </bvar>
        <condition>
          <apply>
            <in/>
            <ci> A </ci>
            <ci> S </ci>
          </apply>
        </condition>
        <ci> E </ci>
      </apply>
      </apply>
      <ci> q </ci>
    </apply>
  </apply>
</math>
```

Fig. 3. MathML content markup encoding equation 1. The grouping element is `apply` instead of `mrow`. The empty element `eq` stands for the equality relation, `times` for multiplication (it does not matter if we want an invisible multiplication here), `int` for an integral. Numbers and identifiers are represented with the token elements `cn` and `ci`. The special token `csymbol` can be used to extend the (necessarily finite) set of basic operators and constants that MathML content provides

3.2   Semantics in Content

MathML content markup's aim is that of encoding the "meaning" of mathematical formulae. A possible encoding of equation 1 in content markup is shown in Fig. 3. The basic idea is the same: the formula is encoded depending on its structure, but grouping is always done with respect to an operator or a function that says how the grouped subexpressions relate to each other. Most MathML content elements allow the `definitionURL` attribute whose value is a reference to another document that gives more precise semantics to the element it applies to. In the figure the `definitionURL` attribute is supposed to point at a resource identifying the symbol `ezero` with the electric permittivity of free space.

MathML content markup can be associated with a corresponding fragment of presentation markup by means of the `semantics` element:

```
<semantics>
  MathML Presentation markup of Fig. 2
  <annotation-xml encoding="MathML-Content">
    MathML Content markup of Fig. 3
  </annotation-xml>
</semantics>
```

Since XML documents are trees with labelled nodes, a fragment of an XML document is a subtree easily identifiable either by an explicit *id* on the node, in the form of an attribute, or by a path from the root of the tree. XML provides a linking mechanism that can be exploited to relate fragments of presentation markup to fragments of content markup within the same `semantics` element. In our favorite equation, we can set a link from the `mo` element encoding the contour integral (which we know being a surface integral) to the corresponding `int` element: the `xref` attribute in Fig. 2 and the `id` attribute in Fig. 3 do exactly this. As a result, we can avoid any ambiguity that may come from the use of a particular notation, as long as we can access the related content by means of suitable links.

Even more severe are those situations in which the structure of presentation markup *must be violated* in order to achieve particular rendering effects. Breaking the structure is an irreparable process that prevents in all but the simplest cases to recover any meaning from the encoded formula.

Two typical examples where this occurs in classical mathematics are the following:

- very long expressions that need to be manually arranged on several lines, as the automatic line-breaking algorithm is not available, or does not produce a satisfactory rendering;
- functions defined "by cases", which do not have a linear, horizontal formatting and their vertical formatting is not completely standardized.

The use of `semantics` allows the preservation of the meaning by providing a parallel encoding of expressions: the presentation markup achieves the required rendering, possibly paying the price of not being completely well-structured; content markup encodes the meaning, and links can relate well-structured fragments of the presentation markup with corresponding fragments of content markup, giving rendering engines and, more generally, any processor of the markup, awareness of the semantics.

4   Formatting Quality

The LATEX user will be concerned about quality of a (visual) rendering engine for MathML, since he comes from a world where quality is "for free." The TEXnician also knows that, in case any difficulty arises, he can count on the low-level TEX commands and macros with which everything is virtually possible (apart from being Turing complete, TEX's primitive commands have been specifically designed for typesetting purposes).

Among the characteristics that made TEX a quality typesetting system since the very beginning were its association with a family of high-quality fonts and the fact that TEX has basically fixed formatting rules. Authors who need to tweak a formula are allowed to do so, and are guaranteed that all TEX implementations will render it identically, if they use the same fonts. MathML, on the other side, has completely different goals: as a Web-oriented language it aims at device (and font) independence; it has to be suitable for rendering in a wide variety of environments ranging from printed paper to computer screens, to pocket-size devices; it has to enable interactive capabilities where feasible. Hence, MathML formatting rules are much looser and the fine points are up to the rendering engine, although MathML does suggest certain conventions. Notwithstanding this, we have shown in a different work [13,14] that it is possible to design and implement a general architecture for MathML formatting that achieves the best formatting available for each context, in particular a formatting quality comparable to that of LATEX where the context allows it.

The basic idea of the architecture is the following: MathML presentation elements correspond to *formatting devices*, each having a specific *formatting semantics*. The formatting engine is made of two components: a set of modules implementing the formatting semantics of the MathML elements and an abstraction of the formatting context that hides all the details regarding the available fonts, their properties, the resolution of the output device and so on.

One of the innovative features of the architecture is that the abstraction of the formatting environment is specifically designed for formatting mathematics, and it can be instantiated to different implementations depending on the platform, the available fonts, and, more generally, the context in which the engine is used. By contrast, common multi-platform applications that render structured documents[4] are based on a very primitive abstraction of the environment, that makes them unsuitable to achieve quality typesetting in every context.

Modules of the formatting engine are such that they interact with each other by means of a simple and narrow interface, allowing the interchange of different

---

[4] Mozilla (`http://www.mozilla.org`) and AbiWord (`http://www.abiword.org`) are two examples.

module implementations for the same formatting device. For instance, a full-featured implementation of the module corresponding to the `mrow` element might provide support for automatic line-breaking,[5] whereas another implementation, which is targeted to embedded system with computational constraints, might ignore the problem thus providing a more light-weight module.

In order to be formatted, the source document is parsed and compiled into a tree of *areas*. Areas describe low-level typographical entities (glyphs, horizontal or vertical spaces) and their geometrical relationships (horizontal or vertical grouping, overlapping). The architecture can be related to TeX and LaTeX as follows:

- areas are close to TeX primitives such as `\hbox` and `\vbox`, used in conjunction with so-called "kerns" and "fillers." Area creation and management are hidden inside the abstraction of the formatting environment. Areas may depend on the available fonts, on their properties, and generally on the formatting context;
- modules are close to LaTeX macros, but they are completely independent of the formatting context as the areas they generate indirectly through the abstraction of the environment are opaque to them. The only property that modules can query about areas is their *bounding box*, that is the extent of the rectangular portion of the output medium they occupy.

An interesting consequence of the architecture's design is that, by a clear separation of modules from areas, we were able to describe the process of formatting mathematics encoded in MathML in a formal way using the language of areas.

The issue of quality could have been quickly addressed by noticing that, thanks to the close correspondence of LaTeX and MathML layout schemata (Table 1), and given the availability of a tool for converting MathML presentation markup into LaTeX markup (this task can be accomplished with an XSLT stylesheet, Sect. 5) the two languages are potentially equivalent in terms of formatting quality. But the transformation of MathML into LaTeX is not loss-less in general, for much of the semantical information which is explicitly encoded in the MathML presentation markup is forgotten in the process. In fact, we can say that with MathML presentation markup it is possible to achieve the same formatting quality achieved with TeX and LaTeX, but with basically no tweaking at the MathML markup level, and by doing this in a context-sensitive, adaptable way.

5   Macros and Transformations

The whole TeX system is based on the concept of *macro*. A macro is a named, possibly parametrized sequence of TeX commands that expands wherever the

---

[5] Support for line-breaking is affected by other MathML elements as well, but it is mainly governed by `mrow`.

name occurs in the document. The macro mechanism of TEX is what ultimately makes the system usable, for typing at the level of the basic commands is really hard. LATEX is "just" a large collection of TEX macros that allow the author to typeset mathematical documents at a relatively high level of abstraction, so that it is not necessary to use basic TEX commands directly (even if some people, including TEX's creator, still prefer to do so). In TEX a macro is introduced with the \def command, which is followed by the name of the macro we want to define, a *pattern* with named holes (the arguments of the macro), and a body that the macro expands to. The arguments may be referred to in the body and are substituted by their value at each point of expansion. In LATEX the user can introduce a macro with \newcommand, with a syntax which is a cleaner and simplified (hence less powerful) version of the one for \def.

```
<xsl:template match="m:apply[*[1][self::m:int]
                             and *[3][self::m:condition]]">
  <m:mrow>
    <m:msub>
      <m:mo> &oint; </m:mo>
      <xsl:apply-templates select="*[3]"/>
    </m:msub>
    <m:mrow>
      <xsl:apply-templates select="*[4]"/>
      <m:mo> &sdot; </m:mo>
      <m:mrow>
        <m:mo> &DifferentialD; </m:mo>
        <xsl:apply-templates select="*[2]"/>
      </m:mrow>
    </m:mrow>
  </m:mrow>
</xsl:template>
```

Fig. 4. An example of XSLT template for converting the integral encoded in Fig. 3 into its corresponding presentation markup in Fig. 2. The template matches against an **apply** element whose first child is the **int** element, and whose third child is a **condition** element. The transformation occurs recursively as the XSLT commands **apply-templates** are executed

MathML does not define any notion of macro. Given that MathML is not meant to be written by hand, it is not clear what advantages could derive from a macro mechanism other than a smaller size of documents. Besides, XML has already a powerful, standard transformation mechanism that goes under the name of XSLT (XSL Transformations[6]). Despite its name, XSLT is very expressive (it is Turing complete, in fact), thus it can be used to perform arbitrary transfor-

---

[6] See http://www.w3.org/TR/xslt, XSL stands for the Extensible Stylesheet Language.

mations on XML (hence MathML) documents, not limited to "styling." XSLT is a tree transformation language that works by pattern matching on the structure of the document tree, transforming a source document into a result document: an XSLT stylesheet is made of a list of templates that match fragments of the source document. With each template a fragment of the result document markup is specified, either by explicit markup or by re-arrangement and composition of the source document. A typical XSLT stylesheet for MathML converts content markup into presentation markup. Fig. 4 shows a simplified XSLT template for converting integrals encoded in content markup into the corresponding presentation markup. The `match` clause is used by the XSLT engine to determine whether there is an instance of the template in the source document. In case of match, the markup included in the body of the template is generated, and the transformation is applied recursively as specified by the `apply-templates` commands.

Despite the fact that the markup in Fig. 3 is basically as verbose as the markup in Fig. 2, we can think of content as a macro version of presentation that captures structural information about a mathematical formula, regardless of any cultural or environmental context about the formula itself (Fig. 5 gives an example of different renderings of the sine function). In this view, XSLT provides a mechanism that is very close to *macro expansion* of content markup into presentation markup. Other approaches based on MathML extensions are also being investigated [12].

Remarkably, XSLT engines are becoming a standard component of the most popular Web browsers, thus enabling on-the-fly transformations of XML documents, including those embedding mathematics. A combination of MathML presentation and content markup together with XSLT stylesheets could be easily exploited to customize mathematical documents on-the-fly, according to the user's environment, culture, and personal preferences. Such transformations can also automate the generation of links connecting freshly created presentation markup with the source content markup, thus relieving authors of mathematical documents from this burden.

6 Conclusions

In the last decade the world of publishing and the World Wide Web have become increasingly involved. This is a one-way transformation process that will eventually lead to a new concept of document and resource, especially in the scientific area. Already now we can touch new digital libraries, electronic journals, online encyclopedias, specialized search engines that have adopted, or are going to adopt, MathML and its related technologies as the basis for the management and processing of mathematical knowledge. An increasing number of research projects are exploring and developing the possibilities of this approach, and the results achieved so far are tangibly promising. We are just a little step away from having interactive, semantic and context sensitive, hyper-linked libraries of scientific documents.

```
<apply>
  <sin/>
  <ci> x </ci>
</apply>
```

(a) 'the sine of x'

```
<mrow>
  <mi> sin </mi>
  <mo> &ApplyFunction; </mo>
  <mi> x </mi>
</mrow>
```

(b) $\sin x$

```
<mrow>
  <mi> sin </mi>
  <mo> &ApplyFunction; </mo>
  <mfenced>
    <mi> x </mi>
  </mfenced>
</mrow>
```

(c) $\sin(x)$

```
<mrow>
  <mi> sh </mi>
  <mo> &ApplyFunction; </mo>
  <mi> x </mi>
</mrow>
```

(d) $\operatorname{sh} x$

Fig. 5. The formula $\sin x$ in MathML content markup (a) and three expansions to presentation markup. Expansion (d) is particularly common in Russian literature.

People will still use LaTeX for a long time in the future because of its relatively simple syntax, that today is known by virtually every author of publications in computer science, in mathematics, in chemistry and certainly in other fields. But here is the key: people will use its syntax, but will tend to forget the underlying system, since the potential of MathML and, in general, XML technologies with respect to document processing, flexibility of information management, and expectations for Web publishing are overwhelming. Because of its nature, MathML is suitable for on-line, interactive and printed paper publishing, and has the potential of achieving LaTeX high-quality typesetting. Powerful transformation technologies, XSLT first of all, allow possibly *loss-less* transformations between MathML and other XML dialects, as well as on-the-fly localization and customization of the markup. MathML is suitable for the communication of content other than presentation, hence providing disambiguous notation, facilitating communication between computer applications, and allowing semantic-sensitive searching operations over digital libraries of mathematics.

References

1. R. Ausbrooks, et.al., *Mathematical Markup Language (MathML)*, Version 2.0, W3C Recommendation, February 2001. Online http://www.w3.org/TR/MathML2
2. B. Beeton, A. Freytag, M. Sargent III, *Unicode Support for Mathematics*, Proposed Draft, Unicode Technical Report #25, 2002. Online http://www.unicode.org/unicode/reports/tr25/
3. D. E. Knuth, *The TeXbook*, Addison-Wesley, Reading, Massachusetts, 1984.

4. D. E. Knuth, *The METAFONTbook*, Addison-Wesley, Reading, Massachusetts, 1986.
5. D. E. Knuth, *Mathematical Typography*, Bullettin of the American Mathematical Society (new series) **1**, pp.337–372, March 1979.
6. D. E. Knuth, `TEXDR.AFT`, in *Digital Typography*, CSLI Lecture Notes Number 78, pp.481–504, 1999.
7. D. E. Knuth, `TEX.ONE`, in *Digital Typography*, CSLI Lecture Notes Number 78, pp.505–532, 1999.
8. D. Halliday, R. Resnick, K. S. Krane, *Physics*, 4th Edition, John Wiley & Sons, Inc., 1992.
9. S. Huerter, I. Rodionov, S. Watt, *Content-Faithful Transformations for MathML*, MathML International Conference 2002, Chicago, Illinois. Online
   `http://www.mathmlconference.org/2002/presentations/huerter/`
10. L. Lamport, *LATEX—A Document Preparation System*, Addison-Wesley, Reading, Massachusetts, 1985.
11. L. Lamport, *LATEX—A Document Preparation System*, 2nd edn. for LATEX 2_ε, Addison-Wesley, Reading, Massachusetts, 1994.
12. W. A. Naylor, S. Watt, *Meta Style Sheets for the Conversion of Mathematical Documents into Multiple Forms*, Proceedings of the Mathematical Knowledge Management, RISC-Linz, Austria, September 2001.
13. L. Padovani, *A Standalone Rendering Engine for MathML*, MathML International Conference 2002, Chicago, Illinois. Online
   `http://www.mathmlconference.org/2002/presentations/padovani/`
14. L. Padovani, *MathML Formatting*, Ph.D. Thesis, Bologna, Italy. To be published in 2003.
15. *The Unicode Standard*, Version 3.0, Addison-Wesley Developers Press, Reading, Massachusetts, 2000. Online
   `http://www.unicode.org/unicode/uni2book/u2.html`
16. U. Vieth, *Math typesetting in TEX: The good, the bad, the ugly*, Proceedings of the EuroTEX'01 Conference.

Problems and Solutions for Markup for Mathematical Examples and Exercises*

Georgi Goguadze[1], Erica Melis[1], Carsten Ullrich[1], and Paul Cairns[2]

[1] DFKI Saarbrücken, D-66123 Saarbrücken, Germany,
{george,melis,cullrich}@ags.uni-sb.de
[2] UCL Interaction Centre, University College London,
26 Bedford Way, London WC1H OAP, UK
p.cairns@ucl.ac.uk

Abstract. This paper reports some deficiencies of the current status of the markup for mathematical documents, OMDoc, and proposes extensions. The observations described arose from trying to represent mathematical knowledge with the goal to present it according to several well-established teaching strategies for mathematics through the learning environment ACTIVEMATH. The main concern here is with examples, exercises, and proofs.

Keywords: knowledge representation, markup for mathematics documents

1 Motivation

For publishing mathematics on the Web OpenMath [6] and MathML[8] representations are being developed, for mathematical symbols and expressions and OMDoc[13] that integrates both, OpenMath and MathML and, in addition, takes care of structural elements of mathematical documents. Such standard representations are necessary, among others, in order to present the same knowledge in different contexts and scenarios without changing the underlying representation.

Naturally, these representations develop over time as more and more applications occur and new requirements built up. Changes or extensions might be required, e.g., when more theorem proving systems want to use the same representation or when authors want to encode documents and present them in a useful but not yet covered style.

The ACTIVEMATH-group collaborates with several authors who emphasize different elements of a mathematical document and want to realize different styles of presentation and teaching which - in principle - is supported by AC-TIVEMATH' separation of representation and presentations. The needs of the various authors are very important for establishing a reasonable standard since

---

* This work has been funded by a project funded by the German Ministry for Education and Reseach (BMBF) and by the EU-project MoWGLI

A. Asperti, B. Buchberger, J.H. Davenport (Eds.): MKM 2003, LNCS 2594, pp. 80–92, 2003.

only if a standard is flexible enough to meet the authors' demands, will the standard be accepted widely. This is because there is evidence that even a simple self-determination of the sequencing of content improves the acceptance [1].

This does not mean that every new idea of an author should change the standard. Although we want to be able to simulate as many useful teaching and learning scenarios as possible and assemble appropriate content with ACTIVEMATH or any other user-adaptive environment, we are determined to use document representations that are encoded in a standard way rather than in a very specific and system-dependent way.

This paper reports presentational needs that cannot be realized using the information included in the current representation standards. These needs require extensions of the representation and we propose some of them.

2   Preliminaries

ACTIVEMATH is a web-based learning environment that generates interactive mathematics courses and feedback in a user-adaptive way and dependent on a chosen learning strategy (scenario) [16]. The user-adaptivity is based on the learner's goals, preferences, capabilities, and mastery of concepts which are stored and updated in a user model. For interactive exercises with computer algebra systems (CASs) we developed a generic interface and feedback mechanism [5]. This code still relies on the language of the CAS in use but the plan is to abstract this language.[1]

2.1   Current OMDoc Markup

OMDoc provides markup elements for basic mathematical items such as axiom, definition, theorem, lemma, corollary, conjecture and proof. Other kinds of textual items such as remarks and elaborative texts are represented by an omtext element that can have different types indicating which kind of textual remark it is. Currently, examples and exercises are represented as separate markup elements of the OMDoc core, although they are educational in nature.

Examples in OMDoc have no particular internal structure. Those examples can be annotated with a relation of type for or against to indicate that the example is a model of the concept or not.

Exercises in OMDoc allow for further structuring of their informal textual content and multiple choice questions are represented by a particular OMDoc element additionally for the representation of so-called multiple choice statements. An exercise element consists of the text of its statement, a possible occurrence of a hint element, zero or more solution elements or zero or more mc elements representing multiple choices. The hint is supposed to be a small remark providing some initial idea for a solution. The solution consists of an OMDoc CMP element (CMPs can contain text and mathematical object representations) and

---

[1] We needed to use the feedback practically before we were able to define a more generic language.

is not further structured. An `mc` element can consist of choice, hint (optional), and answer. This does not allow for more complex constructions (see 3.2 for some examples). The markup for exercises is still subject of research. After the modularization of `OMDoc` in one of the next versions, the `exercise` element is going to migrate from the core `OMDoc` to its pedagogical module[2].

Special attention in `OMDoc` is paid to the representation of proofs. A proof in `OMDoc` is a direct acyclic graph (`DAG`) the nodes of which are the proof steps. Every proof step contains a local claim and a justification by applying an inference rule, referring to a proof method, or employing a subproof. Steps can be represented formally as well as contain textual annotations. Since the `DAG` structure of a proof can not be directly represented by `XML` tree-like structure, the `OMDoc` element `proof` consists of a sequence of proof steps, whose `DAG` structure is given by cross-referencing. Steps are represented by four kinds of elements: `derive` element that specifies the proof step deriving new claim from known ones, `hypothesis` element to specify local assumptions, `conclude` element, reserved for the last step in the proof and `metacomment` that contains textual explanations. These elements have some further structure, including cross-references, references to subproofs or external proof methods.

2.2   Current Educational Metadata in the ACTIVEMATH DTD

According to the needs of a learning environment, mathematical items have to be annotated with additional metadata. The ACTIVEMATH DTD defines a set of metadata for most of the `OMDoc` items as well as some metadata specific for exercises.

Among others, it refines the `relation` element by several types of relations relevant for pedagogical purposes and beyond. The standard values of the attribute *type* of the `relation` are "depends-on" as purely mathematical dependency and "prerequisite-of" as a pedagogical dependency pointing to the items needed for the understanding the current one. Another pedagogical relation is the role of a mathematical item relative to another one. There can be multiple role of one element. For example, a definition *for* one concept can serve as an *example-for* another one.

Some pedagogical metadata describe a learning situation similar to those of Learning Object Metadata (`LOM`). These are the fields of an item (mathematics, computer science, economy, etc.) and its learning context (university first cycle, secondary school, etc.), as well as technical difficulty and abstractness of the content.

Characterizations applicable especially to exercises are the activity type with values "check-question", "make-hypothesis", "prove", "model" and "explore" and pedagogical level with values "knowledge", "comprehension", "application" and "transfer" corresponding to the Bloom's taxonomy of learning goal levels.

---

[2] Communication on `OMDoc` list

3   Experiences with Representing Mathematical Documents

Our current experience with authors designing mathematics courses includes:

- Abstract algebra for university students by Cohen, Cuypers and Sterk [9]
- Calculus book for first year university students by Dahn and Wolter [10]
- Statistics course by Grabowski
- Topology course by Cairns [7]
- Operations research course by Izhutkin
- Calculus course for high school (11th grade) by Schwarz and Voss [18]

In the near future, more authors and institutions from Germany, Great Britain, China, Mexico, Russia, Spain and the United States, will join our effort to present mathematics through ACTIVEMATH.

In this section, we report some problems that occurred when we tried to represent these mathematics courses in the current OMDoc and maintain their characteristic presentation. For these problems we suggest some solutions. So far, the need for extending OMDoc has been most urgent for presenting examples, exercises, and proofs properly.

3.1   Example Representation: Problems and Solutions

In OMDoc examples represent concrete mathematical structures, rather than examples in an educational sense (which overlap but have a much wider range). They do not possess internal structure allowing for applying advanced pedagogical strategies.

Examples occur in educational material and in other types of mathematical documents. Depending on the teaching strategy, examples are introduced before a general property ('American style') or after ('German style'). This sequencing can be performed by the ACTIVEMATH course generator without any need to extend the knowledge representation. However, there are more subtle style requirements that a course generator cannot realize without an additional markup.

For instance, the school calculus course [18] is built mostly on examples. It illustrates a concept by an example before and after introducing it. Another strategy in this course is to choose a number of concrete examples and then develop them by changing parameters and extending them to newly introduced concepts. This produces a hierarchy of examples developing in parallel to the conceptual content. This is just one instance of quite a strong use of examples in teaching and beyond.

There are other ways in which examples are used that are common to many mathematical texts and ACTIVEMATH should be able to reflect these common practices. The following lists some of the ways in which examples are used:

1. A particular example may be used as motivation for a theory. For example, in [17], an example of solving simultaneous linear equations is used to motivate the definition of elementary row operations on matrices. The goal of other examples is to illustrate the concept.

2. Examples can have considerable structure, they can be miniature theories themselves. For pedagogical reasons, some authors emphasize this structure. They explicitly provide a problem description and its (possibly worked-out) solution.
3. A single example can be used many times to illustrate many different and various aspects and properties. This can be particularly evident, when an example is used as a prototypical structure for a theory, for instance, in the use of $[0,1]^\omega$ as the prototypical Hilbert Space [4].
4. An example can be constructed over several sub-examples. In [10] the concept of linear function is introduced by giving several examples that are continued when elaborating further properties of linear functions. In [3], the example of the exponential function as a power series is built up in three examples over two chapters.

In order to decide which example is better suited for being a motivation the author would have to provide additional metadata, since this can not be deduced automatically. The problem addressed in 1 can be solved by the *role*-dependency metadata mentioned in section 2.2. In the above example, the relation can be of a type *motivation-for* and contain a reference to the concept "elementary row operations on matrices".

Using a single example to illustrate different properties, the problem addressed in 3 of our list, requires the use of multiple relations of type *for*. Therefore, we have changed the `for` attribute of an `OMDoc` element `example` to an attribute of element `relation` in metadata that can contain more than one reference elements as children.

In `OMDoc` an example element is connected to the concept it is a model for. There is no connection between examples using the same concrete structure if they are bound to different concepts. Therefore, something is missing for use of pedagogical strategies that successively consider the same structure assigning more and more properties to it.

In order to provide more facilities for the reuse of examples additional structure is needed. We propose to partition an example into three subelements: an `situation description` (SD) containing a description of mathematical objects (structures, formulas, terms) and relations between them, considered in this example, one or more `property assignments` (PA), and every property assignment may have one or more `solutions` (SOL).

In addition we propose to connect examples that use the same object description by a new type of relation called *same-situation* provided in the metadata of the example. In this way an object descriptions can be introduced only once and then retrieved automatically. One could also avoid introducing a new type of relation in metadata. In order to reuse the existing situation description it would be sufficient to provide a reference to it inside the body of an example. But the goal of our annotation is to keep all the information necessary for course generation within the metadata of the item. This way there will be no reason to parse the content of the example during the course generation.

The proposed complex structure is not obligatory, so the author can also write the content of his example just as a `CMP` element. The new structure can be described in a simplified `DTD`-style as follows:

```
Example := (CMP|(SD?,PA+))
SD      := (CMP)
PA      := (CMP,SOL*)
```

(The ?-mark here means zero or one, + means one or more, * means zero or more.)

Consider the following example:

- *Let's take a look at the set of real numbers with the addition operation.* – situation description
- *This structure is a monoid.* – assignment of a property
- *Indeed, the addition operation is associative and possesses a unit – the number 0.* – Proof, i.e., problem solution

It has an internal structure that can be emphasized according to the wish of author, as asked in 2. This example can be continued - the more notions are introduced, the more properties can be assigned to the situation description, i.e. the real numbers with addition operation. By providing the *same-situation* relation other examples can reuse the same object description and enable new pedagogical strategies that rely on developing examples.

3.2 Exercise Representation: Problems and Solutions

Historically, multiple choice questions (MCQ) were the first and only explicit type of exercises in OMDoc. In the light of our experience this is not a valid decision because MCQs are just one of several types of exercises occurring in mathematics courses, even a rather untypical one. The main reason for its use in on-line mathematics courses is the simplicity to understand and evaluate the user's input. In order to make an MCQ a valuable learning and assessment source anyway, carefully designed questions and sometimes more structure is required.

Just as examples can serve several purposes in a text, so too can exercises. Of course, the emphasis is different. Whereas examples are illustrative or elaborative, exercises are intended to help the reader develop a deeper understanding through actively solving problems rather than just reading about them. Nevertheless, the uses of exercises in mathematical texts have much in common with the use of examples:

1. A particular exercise can serve multiple purposes, just as an example. For instance, it can be used to motivate theory by implicitly employing ideas that will later be defined in their own right. This problem can be solved in the same way as for examples.
2. An exercise can have considerable structure, for instance, it may consist of many sub-exercises. In fact this is an extremely common structure in exercises so that the reader is effectively led to the final solution rather than being faced with the entire problem from the outset.
3. Exercises may actually develop an entire piece of theory. Willard [19] has many exercises on topological groups that together build up a substantial theory. However, nowhere else in the theorems or definitions in the book

are topological groups mentioned. This case will be handled analogously to examples by providing the relations of the type *same-situation* between corresponding exercises.

Similar to examples, we propose to structure the internal representation of exercises into subelements of the following types:

- `situation-description` (SD) describing the initial state of the problem. This means description of some given objects and relations between them.
- `problem-statement` (PS) containing the statement of the problem to be solved. This can be an assignment of a property, calculation or construction of some new structures. This element can possibly contain a `hint`, `interactive actions` and `solutions`. Multiple `problem statements` for dividing the exercise into sub-exercises can be provided.
- `interactive-action` (IA) serve as an abstraction of the interactive parts of an exercise. Every action consists of an `EXCLET` and one or more `feedback` elements.
- The code of the interactive pieces itself is contained within the subelement `EXCLET`. For example, it can contain an `MCQ`, a table or a text with blanks to fill-in, a call of an external system such as `CAS` or `planner`, or an applet in which the user has to analyze (modify) some graphical objects. Abstractly, the execution of an `EXCLET` should return a certain result which is then compared with the finite number of `conditions` of the `feedback` elements in order to select the appropriate feedback.
- `feedback` (FB) consisting of the `condition` that has to be fulfilled for the feedback to be given, and two kinds of `feedback`: `system messages` sent to interested system components, for instance the user model, and `user messages` to be provided to the user. It can also contain another problem statement that enables nesting of interactive actions.
- `condition` to be satisfied for the feedback to be applied is a string that is compared to the result of an `EXCLET`.
- `system message` (SYS-MES) for providing information to the user model or other components of the learning environment on the result of user interaction
- `user messages` (USR-MES) for the communication with the user,
- `solution` (SOL) containing a solution of the `problem statement`. Solutions share several structure properties with proofs. As soon as a suitable representation for proofs is specified, it can be also used for worked-out solutions.

The new structure of exercises is shown by the following schema:

```
Exercise := (CMP|(SD?,PS+))
SD        := (CMP)
PS        := (CMP,HINT?,IA*,SOL*)
HINT      := (CMP)
IA        := (CMP,EXCLET,FB*)
FB        := (condition,SYS-MES*,USR-MES,PS*)
USR-MES   := (CMP)
```

Let's take a closer look at the EXCLET element. This element serves as an abstraction for all types of interactive steps. All interactive exercises have in common that after the execution of each of their steps the result of the step is evaluated and some information has to be provided to system components such as the user model and feedback to the user himself. The outcome of the interactive step is compared to the condition of the feedback elements. The matching feedback element is executed, i.e., its messages are delivered and its other child elements (such as another problem statement) are presented.

The following schema shows the structure of a simple MCQ as an instance of an EXCLET. It is, in fact, the simplest kind of an interactive step. It consists of a number of choices (CH), which contain the result later compared with the condition of each of the feedback elements.

```
MCQ := (CH*)
CH  := (CMP, result)
```

Note that this structure of MCQ differs from the current OMDoc representation. Choices and the process of their selection is a part of an interactive action and the assignment of feedback to the result is happening outside of the EXCLET.

Let us show how the above representation covers exercises with sub-exercises as well as nested exercises. Table 1 to 3 show different kinds of MCQs that are covered and Figure 1 schematically illustrates their structure. While Table 1 contains an ordinary, non-nested MCQ, Tables 2 show a nested MCQ and and Table 3 an MCQ with subtasks using an external theorem prover.

Given a set of natural numbers with an addition operation. Which of the following structures is it?

Table 1. An ordinary MCQ

Choice:	Answer:
Group	No.
Monoid	Wrong.
None of these	Correct.

The first diagram on Figure 1 illustrates the exercise representation that consists of a situation description, problem statement and one interactive action. This action consists of an EXCLET (with an MCQ inside) and three feedback elements corresponding to three possible values of the result of an EXCLET. The second feedback in the second diagram gives raise to a subproblem with another interactive action. The third diagram is nesting different kinds of interactive actions. At every interactive action the information can be provided to the system components, and to the user.

88   G. Goguadze et al.

Given a set of natural numbers with an addition operation and zero. Which of the following structures is it?

Table 2. A nested MCQ

Choice:	Answer:		
Monoid	Correct		
Group	No, because	Choice:	Answer:
		It has no unit	Wrong
		No inverses	Correct

Given a set of natural numbers with an addition operation. Is it :

Table 3. An exercise using an external theorem prover

Choice:	Answer:
A group?	Click here to prove it with ΩMEGA
A monoid?	Click here to prove it with ΩMEGA

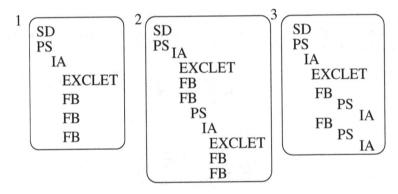

Fig. 1. Exercise structure according to the proposed representation

Interactive Exercises Using External Systems. Currently, ACTIVEMATH offers interactive exercises for which an external system such as CAS or a proof planner can be used to execute certain steps and/or to evaluate the user's input. The general scheme for encoding those exercises includes startup, evaluation, example of which is shown in Figure 2, and shutdown code. These are capsuled in an EXCLET element.

```
<eval><command>evalSilent</command>
  <param>if (Inv=I1) then erg:=1 end_if</param></eval> ...
<eval><command>eval</command>
  <param>
    case erg
      of 1 do print(Unquoted, "This answer is correct!");break
        ...
      of 4 do print(Unquoted, "You must use the variable
                              Inv for the result.");break
    end_case
  </param> </eval>
```

Fig. 2. A paste of an evaluation code of a CAS exercise

The startup is the initialization part and typically contains load instructions. The evaluation part is executed after a learner's input. Essentially, the eval-code represents an evaluation tree. It checks the validity of a user's action and may provide a feedback.

At the moment, the content of these markup elements consist of native code of a particular CAS. As shown in the Figure 2, the feedback is encoded using the syntax of the CAS. The same holds for the rest of the code. The content of the exercise is mixed with the processing instructions of the particular CAS, and hence, the code is not reusable by another external system.

Using the structure for interactive actions, proposed above, one can reach more generic representation of such exercises. For example, the user messages are separated from the code of CAS and represented in a reusable way. Interactive actions are more structured and allow for mixing steps done by different external systems. Some problems, however, are still open.

Problems. One of the obvious problems is the representation of the formula/expression input in the proprietary format of the CAS which is only due to insufficient or buggy phrase-books for the CASs. This is, however, a practical problem only and no longer a research problem. The following lists problems that are not yet fully solved.

 - system-independent encoding of functionalities/commands
 - general encoding of evaluation information (tree) that is the basis for the local feedback and for an evaluation function that serves as input for a user model.

3.3   Proof Representation

Proof is central in modern mathematics. It is the heart of what mathematics is about and yet, in the majority of mathematical texts, has an informal, natural language-like presentation which is difficult to characterize in a way that ACTIVEMATH could currently exploit without a further annotation. This can be seen in the sheer variety of content that a proof can present:

- Proof sketches give an outline of a proof that, whilst not complete, highlight key ideas important for the learner.
- Proof *processes* are really what students need to learn, that is, how to discover/invent proofs for new problems. Yet ironically, the proving process is almost entirely absent from most mathematical texts. With the computer-aided facilities for theorem proving, in particular for proof planning, the picture changes dramatically since they concentrate on the proof *process* and make some information explicit that is only hidden or implicit in minimal proofs. For instance, the collection of constraints during a proof process that eventually serves to instantiate meta-variables, i.e., construct mathematical objects, is not part of the final proof anymore. Typically, for a student such a construction of an object comes out of the blue, maybe not so for a trained mathematician.
- Proofs in traditional mathematical documents contain the final (cleaned up and minimal) proofs only. Much of the information is hidden or implicit in those minimal proofs and could be inferred by an experienced mathematician but not by *any* user.
- Proofs can take the form of calculations, inductions, deductions and discussion. All of these may require different structures to represent them [14, 7].
- Proofs made by humans are usually very high-level and in an informal language [11] that computers cannot use. Proofs generated by computers are usually very low-level and too detailed for humans to use.
- Proofs depend on the notation and theory already developed, be that in the mathematical or logic-calculus sense. The same logical proof may therefore appear very differently in different contexts.
- Proofs can motivate subsequent theory. For example, Euclidean rings are defined as a result of the Euclidean algorithm used to prove that any two numbers have a greatest common divisor [12].

It is a fact that proofs have such rich and diverse communication goals that causes the problem. Any single minimal formalization of a proof structure is likely to omit some of the informal uses for which proofs are used. Proofs could also cause difficulties when translated between the formal contexts of different automated proof systems. For instance, an issue in the MoWGLI project[3] is that the system-independent proof representation should allow for machine-understandability and its reproducibility.

The current OMDoc markup gives a stepwise proof representation which is not suitable for all applications. For example, a hierarchic representation in the form of an expandable proof plan is not possible in OMDoc but would be appreciated by many Web-publishing and learning applications as well as by proof assistants.

First Solutions. The proof process can be presented only via replay in an external system (e.g., proof planner or automated theorem prover).

Proof sketches can be indicated by a value of the metadata verbosity much like other textual elements in the ACTIVEMATH DTD.

---

[3] http://www.mowgli.cs.unibo.it/

More structure can be introduced by providing metadata like *expands-to* connecting pieces with different verbosity or by using Lamport's proof structure that defines a grammar for different elements of a proof [14].

A high-level proof presentation can be achieved as soon as one has richer representation for the proof structure, e.g., by proof planning methods as in [15] will be available.

The machine understandability and reproducibility of proof in different systems requires at least to annotate the proof by its underlying logical foundation and make assumptions explicit that are implicit in a specific proof system.

In order to view a proof as a motivation one can provide a relation of a type *motivation-for* in the metadata containing a reference to the corresponding concepts (as in case of examples and exercises).

4   Conclusion

We have proposed several extensions of the OMDoc markup and of its education-module that is planned for OMDoc 2.2.

Future Plans. We suggest that the metadata be represented in RDF[4] format, because it serves the description and exchange of Web resources and their meta-data. RDF provides a collection of classes (called schema) that are hierarchically organized and it offers extensibility through subclass refinement. This representation will allow authors to define modularized metadata for different applications as well as different sets of metadata of the same module for the same item. In this way other authors could reuse the same items but for their own purposes. The RDF-representation is similar to an object-oriented approach [2] and allows authors to construct new uses for items from existing ones in much the same way as inheritance in object-oriented data structures.

Some Open Questions: We have not solved all the problems we encountered, partly because we do not know a solution yet, partly because there is no agreement on a general standard extension yet. For instance, the following problems are open

- standard language for service system-commands and feedback in order to represent interactive exercises and their local feedback
- standard for proof representation satisfying the needs of formal proof applications as well as Web-publishing and learning applications.

References

1. S. Ainsworth, D. Clarke, and R. Gaizaukas. Using edit distances algorithms to compare alternative approaches to ITS authoring. In S.A. Cerri, G. Gouarderes, and F. Paraguacu, editors, Intelligent Tutoring Systems, 6th International Conference, ITS2002, volume 2363 of LNCS, pages 873-882. Springer-Verlag, 2002.

---

[4] http://www.w3.org/RDF

2. S. Bennett, S. McRobb, R. Farmer, Object-Oriented Systems Analysis and Design using UML, McGraw Hill, 1999
3. K. G. Binmore, Mathematical Analysis: a straightforward approach, second edition, Cambridge University Press, 1982
4. B. Bollobás, Linear Analysis: an introductory course, Cambridge University Press, 1990
5. J. Büdenbender, E. Andres, Adrian Frischauf, G. Goguadze, P. Libbrecht, E. Melis, and C. Ullrich. Using computer algebra systems as cognitive tools. In S.A. Cerri, G. Gouarderes, and F. Paraguacu, editors, 6th International Conference on Intelligent Tutor Systems (ITS-2002), 2363 Lecture Notes in Computer Science, pages 802-810. Springer-Verlag, 2002.
6. O. Caprotti and A. M. Cohen. Draft of the open math standard, Open Math Consortium, http://www.nag.co.uk/projects/OpenMath/omstd/, 1998.
7. P. Cairns, J. Gow, On Dynamically Presenting A Topology Course, Submitted to Annals of Mathematics and Artificial Intelligence (Special issue on Mathematical Knowledge Management) http://www.uclic.ucl.ac.uk/topology/
8. D. Carlisle, P. Ion, R. Miner, and N. Poppelier. Mathematical markup language, version 2.0, 2001. http://www.w3.org/TR/MathML2/.
9. A. Cohen, H. Cuypers, and H. Sterk. Algebra Interactive! Springer-Verlag, 1999.
10. B.I. Dahn and H. Wolter. Analysis Individuell, Springer-Verlag, 2000.
11. J. Harrison, Formalized Mathematics, Math. Universalis, 2, 1996
12. I. N. Herstein, Topics in Algebra, second edition, Wiley, 1975
13. M. Kohlhase, OMDoc: Towards an OpenMath Representation of Mathematical Documents, Seki Report,FR Informatik, Universität des Saarlandes, 2000.
14. L. Lamport, How to write a proof, American Mathematical Monthly, 102(7) p600-608, 1994
15. E. Melis, U. Leron A Proof Presentation Suitable for Teaching Proofs, 9th International Conference on Artificial Intelligence in Education, pages 483-490, 1999.
16. E. Melis, J. Büdenbender, E. Andres, Adrian Frischauf, G. Goguadze, P. Libbrecht, M. Pollet, and C. Ullrich. Activemath: A generic and adaptive web-based learning environment, Artificial Intelligence and Education, 12(4), 2001.
17. A. O. Morris, Linear Algebra: an introduction, second edition, van Nostrand Reinhold, 1982
18. A. Schwarz, M. Voss, Universität des Saarlandes, Multimedia im Mathematik-Unterricht, Copyright ATMedia GmbH, 1998,1999
19. S. Willard, General Topology, Addison Wesley, 1970

An Annotated Corpus and a Grammar Model of Theorem Description

Yusuke Baba[1] and Masakazu Suzuki[2]

[1] Graduate School of Mathematics, Kyushu University
[2] Faculty of Mathematics, Kyushu University
6-10-1 Hakozaki, Higashi-ku, Fukuoka, 812-8581 Japan
{ma201040,suzuki}@math.kyushu-u.ac.jp

Abstract. Digitizing documents is becoming increasingly popular in various fields, and training computers to understand the contents of digitized documents is of growing interest. Since the early 90's, research of natural language processing using large annotated corpora such as the Penn TreeBank has developed. Applying the methods of corpus-based research, we built a syntactically annotated corpus of theorem descriptions, using a book of set theory, and extracted a grammar model of theorems from the obtained corpus, as the first step to understanding mathematical documents by computer.

1 Introduction

In recent years, digitizing documents has become increasingly popular in various fields, for example, in mathematics[1][2][3][4]. In connection with this movement, understanding the contents of digitized documents by computer is of growing interest[5][6].

The technology of understanding documents is applicable to useful systems such as machine translation, summarization and search. In mathematics, there are also particular application ideas of application e.g. the a system that translates descriptions of a proof into a language of proof-checker.

The fundamental technology of understanding documents by computer is parsing. The effectiveness of the above-mentioned systems depends heavily on the accuracy of the parser. Generally, parsing is complicated by the characteristics of natural language such as ambiguity, omission and inversion. On the other hand, since mathematical documents are written in logical and precise expressions, we can expect that the typical methods of parsing for natural language give more accurate results for mathematical documents.

The availability of large, syntactically annotated corpora such as the University of Pennsylvania Tree Bank(Penn TreeBank,[7]) lead to rapid developments in the field of natural language processing. Sekine et al.[8] extracted rules of grammar from the Penn TreeBank and released a parser of English, the "Apple Pie Parser[9]", using the grammar.

To extract a grammar from the Penn TreeBank, the first approach of Sekine et al. is based on the assumption that the corpus covers most of the possible

A. Asperti, B. Buchberger, J.H. Davenport (Eds.): MKM 2003, LNCS 2594, pp. 93–104, 2003.

sentence structures in a domain. The outline of their idea follows. They assign parts-of-speech to an input sentence using a tagger, and then simply search for the same sequence of parts-of-speech in the corpus. The structure of the matched sequence is the output of their parser.

However, it turned out that this strategy was not practical. Only 4.7% of sentences in the corpus had the same structure as another sentence in the corpus. As a result, they applied the idea to not only sentences but also noun phrases, and extracted rules of grammar about S(sentence) and NP(noun phrase). 77.2% and 98.1% of S and NP structures have the same structure in the corpus, respectively [8].

If there were a large corpus covering most of the possible sentence structures in a domain, the method [8] would be more effective. While building a large corpus of natural language costs a great deal of labor and time, a corpus of theorem descriptions covering all the possible structures looks buildable with comparative ease because theorems have a limited vocabulary and consist of many idiomatic expressions. For this reason, we apply the corpus-based methods to process theorem descriptions.

We built a simply annotated corpus of theorem descriptions using a book of set theory, and extracted a context-free grammar with only three non-terminals from the corpus. The constructed corpus included about 100 instances, and the grammar that was extracted from the corpus had 141 generation rules. In order to evaluate the descriptive power of the obtained grammar, we performed an experiment as described in section 4.

2   Building a Corpus

2.1   Preliminaries

We used a book of set theory[10] to collect samples of theorem descriptions as the source of the corpus.

To build an annotated corpus of theorems, we used 28 categories of Parts-of-Speech(POS) symbols, and 3 categories of phrase and clause symbols.

We processed the theorem descriptions as a sequence of words. In our corpus, a formula is simply a word.

2.2   Sentence Structure

To express the structure of theorem descriptions in the corpus, we defined a symbol of phrase(**NP**) and two symbols of clause(**S, IFC**), as follows.

Proposition-Clause(S)

S is a string of words which expresses a proposition. **S** can include any other propositions in itself. Naturally, a theorem description is described by S.

Examples of S

- If R is a strict order on X , then S is a non-strict order on X.
- $(x_1, x_2) = (y_1, y_2)$ if and only if $x_1 = x_2$ and $y_1 = y_2$.
- X is an open set
- $V_\alpha \subset V_\beta$

If-Clause(IFC)

IFC is a string of words which expresses the assumption of a proposition.

Examples of IFC

- **If $|\mathbf{X}| \leq |\mathbf{Y}|$ and $|\mathbf{Y}| \leq |\mathbf{X}|$**, then $|X| = |Y|$.
- **In ZFC**, every vector space has a basis.

Noun-Phrase(NP)

NP is a string of words which can be processed like a noun.

Examples of NP

- "any non-zero ordinal"
- "the set $P = \{R(x) : x \in X\}$ of R-classes"
- "$P = \{R(x) : x \in X\}$"
- "X"

2.3   Classification of Parts-of-Speech

Referring to the Penn TreeBank which has 48 categories of POS symbols[7], we use 7 original categories and 21 typical categories of POS on the grounds that the target domain of our corpus is only theorems. We defined original categories and typical categories of POS so that all the words that appeared in the samples would be classified naturally.

(a) Original Parts-of-Speech

Theorem descriptions have many idiomatic expressions. For example, out of the 96 instances in the samples, 50 theorems(52.1%) have the structure:
"**If** [*proposition*], **then** [*proposition*]" or "**Let** [*proposition*], **then** [*proposition*]."
11 theorems(11.5%) have the structure: "[*proposition*] **if and only if** [*proposition*]." The words 'if' and 'let' are generally classified as a conjunction and a verb respectively, and 'if and only if' is generally partitioned into 'if' 'and' 'only' 'if', in natural language processing. However, since these words have particular meaning in theorem descriptions, they should be given special treatment. In this way, we defined 7 original categories of POS referring to empirical knowledge of mathematics and rough characteristics obtained from the samples:
(1) Proposition-Conjunction(PPC),
(2) Assumption(IF),

(3) **Existence(EX)**,
(4) **Restriction(RST)**,
(5) **Number(NUM)**,
(6) **Explain-Conjunction(EXPC)**,
(7) **Formula(FML)**.
We give detailed explanations for these categories below.

(1) Proposition-Conjunction(PPC)

PPC is used between two propositions, and expresses a relationship between the propositions.

Examples of PPC

- if and only if
- implies

Examples of Appearances in Samples

- A set is most countable **if and only if** it is finite or countable.
- $P(\alpha)$ **implies** $P(s(\alpha))$ for any cardinal α.

(2) Assumption(IF)

IF is a word used in a description which expresses an assumption of the theorem (or a larger proposition which includes it).

Examples of IF

- if
- let
- suppose that
- assume that

Examples of Appearances in Samples

- **If** P is a partition of X, then $R = \{(x, y) : x, y \in p$ for some $p \in P\}$ is an equivalence relation on X.
- **Let** f be a function from X to Y. Then $\text{Im}(f)$ is a subset of Y.

(3) Existence(EX)

EX is used to express the existence of something in a mathematical sense.

Examples of EX

- there is
- there exists

Examples of Appearances in Samples

- **there is** a bijective function from X to Y
- **there exists** a positive integer n

(4) Restriction(RST)

RST is used in order to impose restrictions on **NP**, in a mathematical sense.

Examples of RST

- such that
- satisfying

Examples of Appearances in Samples

- There is no set S **such that** $x \in S$ if and only if $x \notin x$.
- there is a set $Z \in X$ **such that** $p(Z) = Z$

(5) Number(NUM)

In mathematics there are various quantities, not only one, two, three, $\cdots$, but also 'only one', 'all', 'infinite' $\cdots$, which have particular meanings. **NUM** expresses such a numerical quantity in a mathematical sense.

Examples of NUM

- a
- unique
- any
- only one
- no

Examples of Appearances in Samples

- **a** well-ordered set
- For **any** formula ϕ, $\cdots$
- There is **no** set S such that $x \in S$ if and only if $x \notin x$.

(6) Explain-Conjunction(EXPC)

EXPC is used between propositions, and expresses an explanation or a translation.

Examples of EXPC

- in other words
- that is

Table 1. Typical POS

Symbol	Category	Examples of members
NN	Noun	set, relation, function, $\cdots$
PR	Pronoun	it, we, following, $\cdots$
RPR	Relative Pronoun	what, which, that, whose, $\cdots$
JJ	Adjective	countable, empty, same, $\cdots$
BE	be	is, are, be, $\cdots$
VB	Verb	have, lie, $\cdots$
VBG	Verb, gerund/present participle	subsutituting, asserting, $\cdots$
VBN	Verb, past participle	based, denoted, $\cdots$
MD	Modal	may, can, $\cdots$
RB	Adverb	mutually, moreover, totally, $\cdots$
CNJ	Conjunction	and, but, or, as, $\cdots$
IN	Preposition	on, of, in, from, to, $\cdots$
DT	Determiner	the, $\cdots$
LPAR	Left bracket character	(, {, [
RPAR	Right bracket character	), },]
LOQ	Left open quote	' -and- "
RCQ	Right close quote	' -and- "
SFP	Sentence-final punctuation	.
COM	Comma	,
COL	Colon, Semi-colon	: -and- ;
SYM	Other Symbols	-, *, $\cdots$

Example of Appearances in Samples

– For any set X, there is an injection from X to PX but no bijection between these sets; **that is,** $|X| < |PX|$.

(7) Formula(FML)

FML expresses a mathematical formula. We process a formula as a word.

Examples of FML

– $p : PX \to PX$
– $f[A] = \{f(a) : a \in A\}$
– $(x_1, x_2) = (y_1, y_2)$
– X
– n

(b) Typical Parts-of-Speech

In addition to the Original POS, we defined 21 categories of POS that are generally used in natural language processing. Table 1 is a list of these Typical POS.

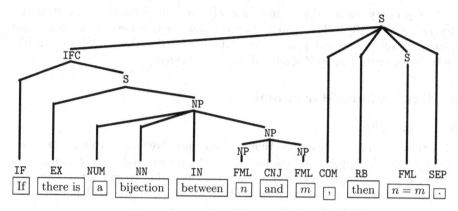

Fig. 1. Example of expression of sentence structure

Table 2. Examples of Members of the Corpus

a: Character String of the Theorem Description
b: Partition of String into Words
c: Sequence of POS
d: Structure of Sentence
   (the numbers 'n' appearing in the expression
   represent the 'n-th' member of a sequence of POS)

a	$(x_1, x_2) = (y_1, y_2)$ if and only if $x_1 = x_2$ and $y_1 = y_2$.
b	[$(x_1, x_2) = (y_1, y_2)$] [if and only if] [$x_1 = x_2$] [and] [$y_1 = y_2$] [.]
c	FML PPC FML CNJ FML SFP
d	(S (S (S 1) 2 (S (S 3) 4 (S 5))) 6)

a	There is no set S such that $x \in S$ if and only if $x \notin x$.
b	[There is] [no] [set] [S] [such that] [$x \in S$] [if and only if] [$x \notin x$] [.]
c	EX NUM NN FML RST FML PPC FML SFP
d	(S (S 1 (NP 2 3 (NP 4) 5 (S (S 6) 7 (S 8)))) 9)

a	The union of at most countably many at most countable sets is at most countable.
b	[The] [union] [of] [at most] [countably] [many] [at most] [countable] [sets] [is] [at most] [countable] [.]
c	DT NN IN RB RB JJ RB JJ NN BE RB JJ SFP
d	(S (S (NP 1 2 3 (NP 4 5 6 (NP 7 8 9))) 10 11 12)13)

a	If there is an injective function from X to Y , and $X \neq \phi$, then there is a surjective function from Y to X.
b	[If] [there is] [an] [injective] [function] [from] [X] [to] [Y] [,] [and] [$X \neq \phi$] [,] [then] [there is] [a] [surjective] [function] [from] [Y] [to] [X] [.]
c	IF EX NUM JJ NN IN FML IN FML COM CNJ FML COM RB EX NUM JJ NN IN FML IN FML SFP
d	(S (IFC 1(S 2(NP 3 4 5 6 (NP 7) 8 (NP 9))10 11 (S 12)) 13 14(S 15(NP16 17 18 19 (NP 20) 21 (NP 22)))23)

Using **S**, **IFC**, **NP** and POS, we assigned the sentence structure for each member of the samples. The structure of a sentence can be expressed as a tree. An example of the expression of a sentence structure is shown in Fig. 1. In the Figure, a box contains those parts of a sentence processed as a word. IF, EX, $\cdots$ are POS. Symbols of phrase and clause are assigned to nodes of the structure tree.

Our primary corpus has 96 instances of theorem descriptions which are all the theorems in [10]. Table 2 shows some examples of members of the corpus. An instance of the corpus has 4 categories of data (a,b,c,d in the Table). Practically, we built the corpus in an XML file using tag hierarchy.

3   Extraction of Grammar

3.1   Algorithm

The structure tree of the corpus that we built has three non-terminal symbols : S, IFC, NP. We applied the method [8] to the corpus and three symbols, and extracted generation rules of a grammar. The algorithm for extracting the rules is as follows. Let G be the set of rules of the grammar to be extracted.
Repeat step1. and step2. for all the trees of the corpus.

The Algorithm of Extraction of Rules

Let G be the set of rules of the grammar to be extracted.
Repeat step1. and step2. for all the trees of the corpus.

step 1.
  For the root and all the nodes of the tree,
  make a generation rule of the grammar such that
  the Left-hand-side of the rule is
      the symbol of the node or root, and
  the Right-hand-side of the rule is
      the left-to-right sequence of the symbols of the children of the node or root.

step 2.
  Add the rules made in step1. to G
      unless any rule is already in G.

Ex. The Rules obtained from Fig. 1 are as follows.

- S -> IFC COM RB S SFP
- S -> EX NP
- S -> FML
- IFC -> IF S
- NP -> NUM NN IN NP
- NP -> NP CNJ NP
- NP -> FML

3.2   The Obtained Grammar

Using Algorithm 3.1, we extracted a grammar which has 141 generation rules, as the model of theorem description. The obtained rules consist of 48 S-rules, 17 IFC-rules, and 76 NP-rules. Table 3 shows a selection of the obtained rules. Since the grammar has only 3 non-terminals, some rules have a long right-hand-side.

Table 3. A selection of the obtained rules of grammar

S-rules (Total 48)
S -> EX NP
S -> FML
S -> IFC SFP RB S
S -> IN NUM NP COM S
S -> NP BE JJ
S -> NP BE NP
S -> NP MD RB VB NP
S -> NP VB
S -> PR BE JJ CNJ JJ
S -> PR BE NP LPAR PR BE COM JJ RPAR
S -> S CNJ S
S -> S COL EXPC COM NP S
. . .

IFC-rules (Total 17)
IFC -> IF S
IFC -> IF S COM IF S
IFC -> IF S SFP COM IF S
IFC -> IN NP
IFC -> IN NUM NP COM IF S
. . .

N-rules (Total 76)
NP -> DT JJ NN
NP -> DT NN IN JJ NN
NP -> DT NN IN NN IN NP IN NP
NP -> DT NN IN NN IN NP RPR VB IN NUM DT NN NP
NP -> NN NN
NP -> NP CNJ NP
NP -> NP CNJ PR NP
NP -> NUM JJ CNJ JJ NN
NP -> NUM JJ JJ NN IN NN RB NP
NP -> NUM JJ NN
NP -> RB JJ NN
NP -> RB RB JJ NP
. . .

4   Experiment

To evaluate the descriptive power of the grammar obtained in section 3(let G be the grammar), we used 100 theorems collected from two books A[11] and B[12] which are written about Galois theory and model theory.

We assigned the correct structure of **S**, **IFC**, **NP** and POS to each theorem, and extracted generation rules using algorithm sec:al. If all the rules obtained from a theorem are elements of G, then the theorem can be described by G. We checked whether or not each theorem is described by G. Additionally, we expanded G into G' by adding the rules which are not in G whenever they appear, and we then checked the descriptivity of G'.

Firstly, we experimented using 50 theorems of book A(Step-1). Secondly we used 50 theorems of book B and G' obtained from Step-1(Step-2). Table 4 shows the results of the experiments. "G-describable" means that the theorem can be described by G. From the table, we can see that G describes only 10% of new theorems. Using the expanded grammar G' did not produce a significant increase in descriptive power.

We can consider that one factor influencing the results is a simplistic structure model. Because our structure has only 3 non-terminals, obtained rules have greater diversity in their right-hand-sides. Moreover, the fact that the total number of rules of G' increased at a constant rate means that the test corpus is too small.

Therefore in order to improve the system, we have to introduce other structures like verb-phrase or adjective-phrase, and expand our corpus.

Table 4. Experimental Results

	G-describable	G'-describable	Total number of rules of G' after Step
Step-1	5/50	11/50	223
Step-2	5/50	11/50	316

5   The Corpus Operation System

To build a large, syntactically-annotated corpus of mathematical text is troublesome work, and very time consuming. In order to make the work easier, we are designing a system that assists with many operations regarding using the corpus.

Fig. 2 is an image of the system. Users need only assign the correct structure of a sentence via a user-interface, and they can see the sentence structure being edited in real time as a tree. Other operations: tagging, parsing, expanding the corpus, training the grammar, are all automatic.

6   Conclusion

We collected about 100 theorem descriptions from a book of set theory[10], and built a model for an annotated corpus of theorems.

We extracted a grammar model of theorem description from the corpus, which has 141 generation rules. We also performed an experiment to evaluate the descriptive power of the grammar.

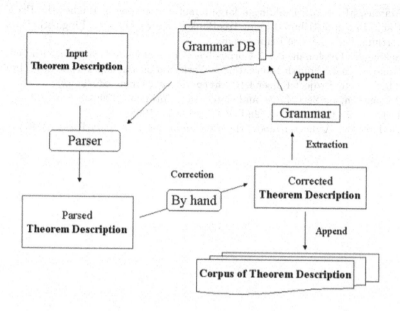

Fig. 2. The Corpus Operatrion System

The results of the experiment indicated that the structure of our test corpus is too simplistic to generate an effective grammar, and the size of the corpus is too small.

As the next step in our work, we will introduce another structure and collect more instances for the corpus. Additionally, we plan to build the semi-automatic corpus operation system in the near future.

References

1. Inoue, K., Miyazaki, R., Suzuki, M.: Optical Recognition of Printed Mathematical Documents. Proceedings of the Third Asian Technology Conference in Mathematics. Springer-Verlag. (1998) 280-289
2. Eto, Y., Suzuki, M.: Mathematical Formula Recognition Using Virtual Link Network. Proceedings of the Sixth International Conference on Document Analysis and Recognition. Seattle. IEEE Computer Society Press. (2001) 430-437
3. Michler, G.: A prototype of a combined digital and retrodigitaized searchable mathematical journal. Lecture Notes in Control and Infomation Sciences. **249** (1999) 219-235
4. Michler, G.: Report on the retrodigitiization project "Archiv der Mathemark". Archiv der Mathemark. **77** (2001) 116-128
5. IPSJ Magazine. vol.41 No.7 July (2000) (in Japanese)
6. IPSJ Magazine. vol.41 No.11 Nov. (2000) (in Japanese)

7. Marcus, M., et al.: Building a large annotated corpus of English:the Penn Tree-Bank. In the distributed Penn TreeBank Project CD-ROM. Linguistic Data Consortium. University of Pennsylvania
8. Sekine, S., Grishman, R.: A Corpus-based Probabilistic Grammar with Only Two Non-terminals. Fourth International Workshop on Parsing Technology. (1995)
9. The Proteus Project Parser URL: `http://nlp.cs.nyu.edu/app/`
10. J Cameron, P.: Sets, Logic and Categories. Springer. (1999)
11. Rotman, J.: Galois theory(2nd ed.). Springer. (1998)
12. Hodges, W.: A shorter model theory. Cambridge University Press. (1997)

A Query Language for a Metadata Framework about Mathematical Resources

Ferruccio Guidi and Irene Schena

Department of Computer Science
Mura Anteo Zamboni 7, 40127 Bologna, ITALY.
{fguidi,schena}@cs.unibo.it

Abstract. Our aim is in the spirit of the Semantic Web: to generate an on-line database of machine-understandable mathematical documents, re-using the same methodology and standard technologies, as RDF, mostly based on the exploitation of metadata to improve smart searching and retrieving. In this paper we describe the syntax and semantics of a query language which gives access to a search engine capable of retrieving resources on the basis of meta-information: the role of metadata consists in filtering the relevant data.

1 Introduction

The mathematical information, owing to its semantically rich *structured* nature, its scalability and inter-operability issues, can be considered as a relevant and interesting test case for the study of content-based systems and the development of the Semantic Web.

The Semantic Web[1] is an extension of the Web in which information is given a well-defined meaning in a machine-processable form, say by means of *metadata*, in order to share, interpret and manipulate it world-wide, enabling and facilitating specific functionalities such as searching and indexing.

Our aim is to generate an on-line database of machine-understandable mathematical documents, re-using the same methodology and standard technologies, as RDF, mostly based on the exploitation of metadata to improve smart searching and retrieving.

Since we are developing our work inside the HELM[2] project [1], which is now integrated with related projects in the framework of the MOWGLI Project[3], we have pointed out and taken into account different kinds of involved domains of application: general mathematical applications, as reviewing databases or on-line libraries, automated mathematics applications, as proof-assistants or proof-checking systems, and educational applications, as learning environments. All these applications require classifying, smart searching and browsing semantically meaningful mathematical information.

---

[1] <http://www.w3.org/2001/sw/>.

[2] Hypertextual Electronic Library of Mathematics, <http://helm.cs.unibo.it/>.

[3] Math On the Web: Get it by Logic and Interfaces, European FET Project IST-2001-33562, <http://mowgli.cs.unibo.it/>.

A. Asperti, B. Buchberger, J.H. Davenport (Eds.): MKM 2003, LNCS 2594, pp. 105–118, 2003.

At this point we have to face two fundamental related issues: the meta-description[4] associated to data and an outfit query language.

We have distinguished at least three different levels of querying mathematical information:

- Querying metadata: filtering relevant data (see MathQL Level 1 in Sect. 2).
- Querying data: taking into account the semantically meaningful content of mathematics (i.e. pattern matching).
- Querying data "up to": processing data on the basis of the *formal* content of mathematics (i.e. isomorphisms, unification, δ-expansion).

The standard format for metadata proposed by W3C is RDF (Resource Description Framework) [4,5]. RDF provides a common model for describing the semantical information on the Web corresponding to different domains of application: RDF allows anyone to express assertions (i.e. statements as subject-predicate-object records) about every resource on the Web.

In this paper we focus on a query language that gives access to a search engine capable of retrieving resources on the basis of meta-information, successfully querying RDF metadata eventually coming from different domains of application.

In particular, besides the main requirements for an RDF query language [10,3, 6], our query language on metadata fulfills two main design goals:

- abstracting from the representation *format* of the queried (meta)data (i.e. XML [9])
- abstracting from the concrete specification of *RDF* but supporting its formal model and characteristics, that are independence, interchange and scalability. Reasoning cannot be based on RDF, but we want an abstract logical model for our query language: this model is given in terms of relationships between resources (see 2).

2   MathQL Level 1

In this section we describe the features of MathQL-1 giving its syntax and semantics. Here are the main guidelines concerning the general language architecture.

- A resource is denoted by a URI reference [7] (or just reference), that is a URI (Uniform Resource Identifier) [7] with an optional fragment identifier, and may have some named attributes describing it. These attributes hold multiple string values and are grouped into sets containing the ones that are meaningful in the same context.
- A query gives a set of distinct resources each with its sets of attributes.
- The basic RDF data model is based on the concept of property: in an RDF statement (s, p, o) the property p plays the central role of the predicate that has a subject s and an object o. This model draws on well-established principles from various data representation communities. An RDF property can be

---

[4] For a detailed description of metadata and RDF schemas for HELM see [6].

thought as an attribute of a resource, corresponding to traditional attribute-value pairs. At the same time, a property can represent relationships between resources, as in an entity-relationship model. In this sense our language is correct and complete with respect to querying an abstract RDF data model. The language allows to build the set of attributed references involved by a property-relationship. In particular we regard a relation between values (that can be resources) and resources as a named set of triples (v, r, B) where v is a string value, r is a reference and B is a set of attributes (see 2.1), providing additional information on the specific connection between v and r (this is the case of a structured property or a structured value of a property).

Moreover the language allows to access any property of a resource by means of functions that we regard as named sets of couples (r, v) where r is a reference and $\{v \mid (r, v)\}$ is its associated multiple string value.

Examples of possible relations and functions can be found in section 3.1 where we discuss an implementation of the language.

– Queries and query results have both a textual syntax and an XML syntax.[5]

2.1   Mathematical Background for MathQL-1 Operational Semantics

We will present MathQL-1 semantics in a natural operational style [8] and we will use a simple type system that includes basic types for strings and booleans, plus some type constructors such as product and exponentiation. $y : Y$ will represent a typing judgement.

Primitive types. The type of boolean values will be denoted by `Boole` and its elements are `T` and `F`.

The type of strings will be denoted by `String` and its elements are the finite sequences of characters. Grammatical productions, represented as strings in angle brackets, will be used to denote the subtype of `String` containing the produced sequences of characters.

Type constructors. $Y * Z$ denotes the product of the types Y and Z whose elements are the ordered pairs (y, z) such that $y : Y$ and $z : Z$. The notation is also extended to a ternary product.

$Y \to Z$ denotes the type of functions from Y to Z and $f\, y$ denotes the application of $f : Y \to Z$ to $y : Y$. Relations over types, such as equality, are seen as functions to `Boole`.

`Listof` Y denotes the type of lists (i.e. ordered finite sequences) over Y. We will use the notation $[y_1, \cdots, y_m]$ for the list whose elements are $y_1, \cdots, y_m$.

`Setof` Y denotes the type of finite sets (i.e. unordered finite sequences without repetitions) over Y. With this constructor we can give a formal meaning to most of the standard set-theoretical notation. For instance we will use the following:

---

[5] In a distributed setting where query engines are implemented as stand-alone components, both queries and query results must travel inside the system so both need to be encoded in clearly defined format with a rigorous semantics.

- $\subseteq$: (Setof Y) $\to$ (Setof Y) $\to$ Boole (infix)
- $\emptyset$: (Setof Y) $\to$ (Setof Y) $\to$ Boole (infix)
- $\sqcup$: (Setof Y) $\to$ (Setof Y) $\to$ (Setof Y) (the disjoint union, infix)

$U \emptyset W$ means $(\exists u \in U)\ u \in W$ and expresses the fact that $U \cap W$ is inhabited as a primitive notion, i.e. without mentioning intersection and equality as for $U \cap W \neq \emptyset$, which is equivalent but may be implemented less efficiently in real cases[6]. $U \emptyset W$ is a natural companion of $U \subseteq W$ being its logical dual (recall that $U \subseteq W$ means $(\forall u \in U)\ u \in W$) and is already being used successfully in the context of a constructive (i.e. intuitionistic and predicative) approach to point-free topology[7].

Sets of couples play a central role in our model and in particular we will use:

- Fst : $(Y \times Z) \to Y$ such that Fst $(y, z) = y$.
- Snd : $(Y \times Z) \to Z$ such that Snd $(y, z) = z$.
- with the same notation, if W contains just one couple whose first component is y, then $W(y)$ is the second component of that couple. In the other cases $W(y)$ is not defined. This operator has type (Setof $(Y \times Z)$) $\to Y \to Z$.
- Moreover $W[y \leftarrow z]$ is the set obtained from W removing every couple whose first component is y and adding the couple (y, z). The type of this operator is: (Setof $(Y \times Z)$) $\to Y \to Z \to$ (Setof $(Y \times Z)$).
- Also $U + W$ is the union of two sets of couples in the following sense:

$$U + \emptyset \text{ rewrites to } U$$
$$U + (W \sqcup \{(y, z)\}) \text{ rewrites to } U[y \leftarrow z] + W$$

The last three operators are used to read, write and join association sets, which are sets of couples such that the first components of two different elements are always different. These sets will be exploited to formalize the memories appearing in evaluation contexts.

Now we are able to type the main objects needed in the formalization:

- An attribute name a is of type String.
- A multiple string value V is an object of type $T_0 =$ Setof String.
- A set of attributes A is an association set connecting the attribute names to their values, therefore its type is $T_1 =$ Setof (String $\times T_0$).
- A reference r is an object of type String.
- The type of a collection D of the sets of attributes of a resource is $T_2 =$ Setof T_1.
- A resource is a reference with its sets of attributes, so its type is $T_3 =$ String $\times T_2$.
- A set of resources S is an association set of type $T_4 =$ Setof T_3.
- A path name s is a non-null list of strings with the first component highlighted therefore its type is $T_5 =$ String $\times$ Listof String.

---

[6] As for $\phi \lor \psi$ which may have a more efficient implementation than $\neg(\neg\phi \land \neg\psi)$.

[7] Sambin G., Gebellato S. *A preview of the basic picture: a new perspective on formal topology.* Lecture Notes in Computer Science n.1657 p.194-207. Springer, 1999.

- The sets B in the triples of relations are association sets of type $\mathtt{Setof}(T_5 \times T_0)$.

We will also need some primitive functions that mostly retrieve the informations that an implemented query engine obtains reading its underlying database. These functions are $\mathtt{Relation}$, $\mathtt{Pattern}$, $\mathtt{Property}$, $\mathtt{Unquote}$ and will be explained when needed.

2.2 Textual Syntax and Semantics of MathQL-1 Queries

MathQL-1 query expressions, or simply expressions, fall into three categories.

- Expressions denoting a set of resources: belong to the grammatical production `<mq_set>`. Their semantics is given by the infix evaluating relation $\Downarrow_s$.
- Expressions denoting a boolean condition: are evaluated by the infix relation $\Downarrow_b$ and belong to the grammatical production `<mq_boole>`.
- Expressions denoting a multiple string value: are evaluated by the infix relation $\Downarrow_v$ and belong to the grammatical production `<mq_val>`.

Expressions can contain quoted strings with the following syntax:

```
<string>      ::= ’"’ [ "\" . | ’^ "\’ ]* ’"’
<path>        ::= <string> [ "/" <string> ]*
<string_list> ::= <string> [ "," <string> ]*
```

When these strings are unquoted, the surrounding double quotes are deleted and each character is not escaped (the escape character is the backslash). This operation is formally performed by the function $\mathtt{Unquote}$ of type $\mathtt{String} \rightarrow \mathtt{String}$. Moreover $\mathtt{Name} : $ `<path>` $\rightarrow T_5$ is an helper function that converts a linearized path name in its structured format. Formally $\mathtt{Name}\ (q_0\ /\ q_1\ /\ \cdots\ /\ q_m)$ rewrites to $(\mathtt{Unquote}\ i_0, [\mathtt{Unquote}\ i_1, \cdots, \mathtt{Unquote}\ i_m])$.

Expressions can also contain variables for single resources (rvar), variables for sets of resources, i.e. for query results (svar) and variables for multiple strings values (vvar).

```
<alpha>  ::= [ ’A - Z’ | ’a - z’ | ‘:_’ ]+
<number> ::= [ ’0 - 9’ ]+
<id>     ::= <alpha> [ <alpha> | <number> ]*
<rvar>   ::= "@" <id>
<svar>   ::= "%" <id>
<vvar>   ::= "$" <id>
```

Expressions are evaluated in a context $\Gamma = (\Gamma_s, \Gamma_r, \Gamma_a, \Gamma_v)$ which is a quadruple of association sets which connect svar's to sets of resources, rvar's to resources, rvar's to sets of attributes and vvar's to multiple string values[8]. Therefore the type K of the context Γ is:

---

[8] Γ_a is an auxiliary context used for **ex** and ``dot'' constructions (see below).

$$\texttt{Setof} \ (\texttt{<svar>} \times T_4) \times \texttt{Setof} \ (\texttt{<rvar>} \times T_3) \times$$
$$\times \ \texttt{Setof} \ (\texttt{<rvar>} \times T_1) \times \texttt{Setof} \ (\texttt{<vvar>} \times T_0)$$

and the three evaluating relations are of the following types:

$$\Downarrow_s : (K \times \texttt{<mq\_set>}) \to T_4 \to \texttt{Boole},$$
$$\Downarrow_b : (K \times \texttt{<mq\_boole>}) \to \texttt{Boole} \to \texttt{Boole},$$
$$\Downarrow_v : (K \times \texttt{<mq\_val>}) \to T_0 \to \texttt{Boole}.$$

Expressions denoting a set of resources. These expressions denote queries or subqueries.

```
<refine>    ::= [ "sub" | "super" ]?
<qualifier> ::= [ "inverse" ]? <refine> <path>
<assign>    ::= <vvar> "<-" <path>
<attr_list> ::= [ "attr" <assign> [ "," <assign> ]* ]?
<mq_set>    ::= "ref" <mq_val> | "pattern" <mq_val>
            | <svar> | <rvar> | "(" <mq_set> ")"
            | "relation" <qualifier> <mq_val> <attr_list>
            | "select" <rvar> "in" <mq_set> "where" <mq_boole>
            | <mq_set> [ "union" | "intersect" | "diff" ] <mq_set>
            | "let" <svar> "be" <mq_set> "in" <mq_set>
            | "let" <vvar> "be" <mq_val> "in" <mq_set>
```

intersect, union, diff are left-associative and have decreasing precedence order.

- The `let` constructions introduce svar's and vvar's in the context; formally:

$$\frac{i : \texttt{<svar>} \quad ((\Gamma_s, \Gamma_r, \Gamma_a, \Gamma_v), x_1) \Downarrow_s S_1 \quad ((\Gamma_s[i \leftarrow S_1], \Gamma_r, \Gamma_a, \Gamma_v), x_2) \Downarrow_s S_2}{((\Gamma_s, \Gamma_r, \Gamma_a, \Gamma_v), \text{let } i \text{ be } x_1 \text{ in } x_2) \Downarrow_s S_2}$$

$$\frac{i : \texttt{<vvar>} \quad ((\Gamma_s, \Gamma_r, \Gamma_a, \Gamma_v), x_1) \Downarrow_v V \quad ((\Gamma_s, \Gamma_r, \Gamma_a, \Gamma_v[i \leftarrow V]), x_2) \Downarrow_s S}{((\Gamma_s, \Gamma_r, \Gamma_a, \Gamma_v), \text{let } i \text{ be } x_1 \text{ in } x_2) \Downarrow_s S}$$

- Then we have some constructions for reading variables and grouping:

$$\frac{i : \texttt{<svar>}}{(\Gamma, i) \Downarrow_s \Gamma_s(i)} \qquad \frac{i : \texttt{<rvar>}}{(\Gamma, i) \Downarrow_s \{\Gamma_r(i)\}} \qquad \frac{(\Gamma, x) \Downarrow_s S}{(\Gamma, (x)) \Downarrow_s S}$$

$\Gamma_s(i)$ and $\{\Gamma_r(i)\}$ mean $\emptyset$ if i is not defined and Γ is $(\Gamma_s, \Gamma_r, \Gamma_a, \Gamma_v)$.
- The `union` construction "sums" two sets of resources in the following way:

$$\frac{(\Gamma, x_1) \Downarrow_s S_1 \quad (\Gamma, x_2) \Downarrow_s S_2}{(\Gamma, x_1 \text{ union } x_2) \Downarrow_s S_1 \oplus S_2}$$

where $\oplus$ puts together the sets of attributes belonging to the same resource:

$$1 \quad (S_1 \sqcup \{(r, D_1)\}) \oplus (S_2 \sqcup \{(r, D_2)\}) \text{ rewrites to } S_1 \oplus S_2 \oplus \{(r, D_1 \cup D_2)\}$$
$$2 \quad S_1 \oplus S_2 \text{ rewrites to } S_1 \cup S_2$$

$\oplus$ is defined associative and rule 1 takes precedence over rule 2.

- The `intersect` construction "multiplies" two sets of resources in this way:

$$\frac{(\Gamma, x_1) \Downarrow_s S_1 \quad (\Gamma, x_2) \Downarrow_s S_2}{(\Gamma, x_1 \text{ intersect } x_2) \Downarrow_s S_1 \otimes S_2}$$

where $\otimes$ builds the product of the sets of attributes belonging to the same resource. Here $D_1 \times D_2 = \{A_1 \oplus A_2 \mid A_1 \in D_1, A_2 \in D_2\}$.

$$1 \quad (S_1 \sqcup \{(r, D_1)\}) \otimes (S_2 \sqcup \{(r, D_2)\}) \text{ rewrites to } (S_1 \otimes S_2) \cup \{(r, D_1 \times D_2)\}$$
$$2 \quad S_1 \otimes S_2 \text{ rewrites to } \emptyset$$

Rule 1 takes precedence over rule 2.

- The `diff` construction makes the "difference" of two sets of resources:

$$\frac{(\Gamma, x_1) \Downarrow_s S_1 \quad (\Gamma, x_2) \Downarrow_s S_2}{(\Gamma, x_1 \text{ diff } x_2) \Downarrow_s S_1 \ominus S_2}$$

where $A \ominus B$ gives the resources of A, with their attributes, that don't belong to B (without considering the associated attributes):

$$1 \quad (S_1 \sqcup \{(r, D_1)\}) \ominus (S_2 \sqcup \{(r, D_2)\}) \text{ rewrites to } S_1 \ominus S_2$$
$$2 \quad S_1 \ominus S_2 \text{ rewrites to } S_1$$

Again rule 1 takes precedence over rule 2.

- The `ref` (reference) construction turns a given set of references into a set of resources, each without attributes (i.e makes a coercion between the types T_0 and T_4); formally:

$$\frac{(\Gamma, x) \Downarrow_v V}{(\Gamma, \text{ref } x) \Downarrow_s \{(v, \emptyset) \mid v \in V\}}$$

- The `pattern` construction gives the set of resources whose reference matches at least one of the given POSIX 1003.2-1992 (included in POSIX 1003.1-2001[9]) regular expressions, each without attributes:

$$\frac{(\Gamma, x) \Downarrow_v V}{(\Gamma, \text{pattern } x) \Downarrow_s \{(r, \emptyset) \mid (\exists v \in V) \, r \in \text{Pattern } v\}}$$

where $\text{Pattern } v$ represents the set of references matching the regular expression v, among the ones available in the underlying database.

- The `relation` construction regards RDF properties for building the set of the resources (r) in the specified relation $\{(v, r, B)\}$, its sub-relations (`sub`) or its super-relations (`super`) with some resource in a given set or with multiple string values in general. Each of these have the set of attributes defined by the assignments following the `attr` clause. In this rule D_2 is $\{\text{Assign } B \{q_1, \cdots, q_n\}\}$ and R is $\text{Relation } f$ ($\text{Name } p$):

$$\frac{f : \text{<refine>} \quad p : \text{<path>} \quad (\Gamma, x) \Downarrow_v V \quad q_1 : \text{<assign>} \quad \ldots \quad q_n : \text{<assign>}}{(\Gamma, \text{relation } f \, p \, x \text{ attr } q_1, \cdots, q_n) \Downarrow_s \bigoplus \{\{(r, D)\} \mid (\exists v \in V) \, (v, r, B) \in R\}}$$

[9] <http://www.unix-systems.org/version3/ieee_std.html>.

Assign builds a set of (eventually compound) attributes whose values are fetched from B using the correspondences given by $q_1, \cdots, q_n$. Formally if $i : $ <vvar> and $p : $ <path>, we define **Assign** $B\ Q = \bigoplus \{(i, B(\texttt{Name } p)) \mid (i \leftarrow p) \in Q\}$. **Relation** $f\ s$ gives the set of triples (R) of the relation s. If $s = (s_0, [s_1, \cdots, s_m])$, **Relation** gives the set of triples of the relation obtained by the composition of the relations $s_0, s_1, \cdots, s_m$. This is the case of a structured property s_0 that has (i.e. is described by) another property s_1 and so on. If f is **sub** or **super**, **Relation** gives also the set of triples of the sub-relations or super-relations of s.

If the **inverse** switch is present, the inverse relation must be used in place of the direct one: while the argument x represents the subject of the property specified by **relation**, on the contrary x represents the object of the property specified by **relation inverse**.

Formally replace R with $R' = \{(r, v, B) \mid (v, r, B) \in R\}$ in the previous rule. Finally note that the vvar's introduced by **relation** differ from the ones introduced by the **let** clause even if they have the same name.

– The **select** construction extracts from a given set of resources the ones meeting a given boolean condition. The condition is tested for each resource of the set and the resources are assigned to a given rvar before the test. Formally we have:

$$\frac{i : \texttt{<rvar>} \quad (\Gamma, x_1) \Downarrow_s S \quad x_2 : \texttt{<mq\_boole>}}{(\Gamma, \text{select } i \text{ in } x_1 \text{ where } x_2) \Downarrow_s \textbf{Select } S\ i\ \Gamma\ x_2}$$

where the **Select** function is defined below.

$$\frac{i : \texttt{<rvar>} \quad x : \texttt{<mq\_boole>}}{\textbf{Select } \emptyset\ i\ \Gamma\ x \text{ rewrites to } \emptyset}$$

$$\frac{i : \texttt{<rvar>} \quad ((\Gamma_s, \Gamma_r[i \leftarrow R], \Gamma_a, \Gamma_v), x) \Downarrow_b \textsf{F}}{\textbf{Select } (S \sqcup \{R\})\ i\ \Gamma\ x \text{ rewrites to } \textbf{Select } S\ i\ \Gamma\ x}$$

$$\frac{i : \texttt{<rvar>} \quad ((\Gamma_s, \Gamma_r[i \leftarrow R], \Gamma_a, \Gamma_v), x) \Downarrow_b \textsf{T}}{\textbf{Select } (S \sqcup \{R\})\ i\ \Gamma\ x \text{ rewrites to } (\textbf{Select } S\ i\ \Gamma\ x) \cup \{R\}}$$

where R is actually a couple of the form (r, D) and $\Gamma = (\Gamma_s, \Gamma_r, \Gamma_a, \Gamma_v)$.

Expressions denoting a boolean condition. These expressions appear in the **where** clause of a **select** construction.

```
<mq_boole> ::= "false" | "true" | "(" <mq_boole> ")"
             | "not" <mq_boole> | "ex" <mq_boole>
             | <mq_boole> [ "and" | "or" ] <mq_boole>
             | <mq_val> [ "sub" | "meet" | "eq" ] <mq_val>
```

and and *or* are left-associative. The precedence (high to low) is: *not, and, or, ex.*

— We have constructions for constants, standard operations and grouping:

$$\frac{}{(\Gamma, \text{false}) \Downarrow_b \text{F}} \quad \frac{}{(\Gamma, \text{true}) \Downarrow_b \text{T}} \quad \frac{(\Gamma, x) \Downarrow_b b}{(\Gamma, (x)) \Downarrow_b b} \quad \frac{(\Gamma, x) \Downarrow_b \text{T}}{(g, \text{not } x) \Downarrow_b \text{F}} \quad \frac{(\Gamma, x) \Downarrow_b \text{F}}{(g, \text{not } x) \Downarrow_b \text{T}}$$

$$\frac{(\Gamma, x_1) \Downarrow_b \text{F}}{(\Gamma, x_1 \text{ and } x_2) \Downarrow_b \text{F}} \quad \frac{(\Gamma, x_1) \Downarrow_b \text{T} \quad (\Gamma, x_2) \Downarrow_b b}{(\Gamma, x_1 \text{ and } x_2) \Downarrow_b b}$$

$$\frac{(\Gamma, x_1) \Downarrow_b \text{T}}{(\Gamma, x_1 \text{ or } x_2) \Downarrow_b \text{T}} \quad \frac{(\Gamma, x_1) \Downarrow_b \text{F} \quad (\Gamma, x_2) \Downarrow_b b}{(\Gamma, x_1 \text{ or } x_2) \Downarrow_b b}$$

The binary operations are evaluated with an early-out (C-style) strategy.
— The **ex** (exists) construction gives access to the sets of attributes associated to the resources in the Γ_r part of the context and does this by loading its Γ_a part, which is used by the `<rvar>.<vvar>` construction (see below).
ex is true if the condition following it is satisfied by at least one pool of attribute sets, one for each resource in the Γ_r part of the context. Formally we have the rule, where $\text{All } \Gamma_r = \{\Delta_a \mid \Delta_a(i) = A \text{ iff } A \in \text{Snd } \Gamma_r(i)\}$ and Δ_a has the type of Γ_a:

$$\frac{}{((\Gamma_s, \Gamma_r, \Gamma_a, \Gamma_v), \text{ex } x) \Downarrow_b ((\exists \Delta_a \in \text{All } \Gamma_r) \, ((\Gamma_s, \Gamma_r, \Gamma_a + \Delta_a, \Gamma_v), x) \Downarrow_b \text{T})}$$

— The **sub**, **meet** and **eq** constructions compare two sets of string values. The comparisons are extensional and the string equality is case-sensitive.

$$\frac{(\Gamma, x_1) \Downarrow_v V_1 \quad (\Gamma, x_2) \Downarrow_v V_2}{(\Gamma, x_1 \text{ sub } x_2) \Downarrow_b (V_1 \subseteq V_2)} \quad \frac{(\Gamma, x_1) \Downarrow_v V_1 \quad (\Gamma, x_2) \Downarrow_v V_2}{(\Gamma, x_1 \text{ meet } x_2) \Downarrow_b (V_1 \between V_2)}$$

$$\frac{(\Gamma, x_1) \Downarrow_v V_1 \quad (\Gamma, x_2) \Downarrow_v V_2}{(\Gamma, x_1 \text{ eq } x_2) \Downarrow_b (V_1 = V_2)}$$

The **eq** operator is introduced because the evaluation of x_1 eq x_2 may be more efficient than that of x_1 sub x_2 and x_2 sub x_1.
As an application of the **sub** and **meet** operators, consider a set of resources denoted by X : `<mq_set>` and a boolean condition denoted by B : `<mq_boole>` and depending on the variable @u : `<rvar>`. The test "is B satisfied for each resource of X?" is expressed by "refof X sub refof select @u in X where B" whereas the dual test "is B satisfied for some resource of X?" is expressed by "refof X meet refof select @u in X where B".
refof makes a coercion between the types T_4 and T_0 (see below).

Expressions denoting a multiple string value. These expressions appear as operands of the **sub**, **meet** and **eq** construction.

```
<mq_val> ::= "{" [ <string_list> ]? "}" | <string>
           | <rvar> "." <vvar> | <vvar> | "(" <mq_val> ")"
           | "refof" <mq_set> | "property" <qualifier> <mq_val>
```

- A set of strings can be given explicitly with these two constructions:

$$\frac{q_1 : \texttt{<string>} \quad \cdots \quad q_m : \texttt{<string>}}{(\Gamma, \{q_1, \cdots, q_m\}) \Downarrow_v \{\texttt{Unquote } q_1, \cdots, \texttt{Unquote } q_m\}} \qquad \frac{q : \texttt{<string>}}{(\Gamma, q) \Downarrow_v \{\texttt{Unquote } q\}}$$

- `refof` (references of) allows to obtain a set of strings gathering the references in a set of resources (i.e. it makes a coercion between the types T_4 and T_0):

$$\frac{(\Gamma, x) \Downarrow_s S}{(\Gamma, \texttt{refof } x) \Downarrow_v \{\texttt{Fst } u \mid u \in S\}}$$

- We have a construction for grouping and two for reading variables: one reads the Γ_a part of the context which is updated by the **ex** clause, while the other read the Γ_v part of the context is updated by the **let** clause for vvar's:

$$\frac{(\Gamma, x) \Downarrow_v V}{(\Gamma, (x)) \Downarrow_v V} \qquad \frac{i : \texttt{<rvar>} \quad j : \texttt{<vvar>}}{(\Gamma, i.p) \Downarrow_v \Gamma_a(i)(j)} \qquad \frac{i : \texttt{<vvar>}}{(\Gamma, i) \Downarrow_v \Gamma_v(i)}$$

where $\Gamma = (\Gamma_s, \Gamma_r, \Gamma_a, \Gamma_v)$. $\Gamma_a(i)(j)$ and $\Gamma_v(i)$ mean $\emptyset$ if i or j are not defined. With the "dot" construction a vvar introduced by a **relation** (i.e. an attribute) can be read only specifying an associated rvar (i.e. a resource) but this restriction offers some advantages: it conforms to the general idea of treating an attribute as an entity which is always related to a resource and allows an unambiguous read of those attributes related to different resources but sharing the same name. Note that the use of the **let** construction may produce unavoidable attribute name collisions in the scope of nested **where** clauses as in the following sample query where ... is a place holder:

```
let %s be relation "..." "..." attr $a <- "a" in
select @u1 in %s where "..." sub refof
    select @u2 in %s where ex @u1.$a sub @u2.$a
```

- Finally a set of strings can be obtained reading the values of a named property, its sub-properties (**sub**) or its super-properties (**super**) of a given set of string arguments (i.e. references subjects of the specified property) with a semantics which is very close to that of the **relation** construction: **property** regards RDF properties to build multiple string values for checking and filtering metadata information.

$$\frac{f : \texttt{<refine>} \quad p : \texttt{<path>} \quad (\Gamma, x) \Downarrow_v V}{(\Gamma, \texttt{property } f \, p \, V) \Downarrow_v \{v \mid (\exists r \in V) \, (r, v) \in Q\}}$$

where $Q = \texttt{Property } f \, (\texttt{Name } p)$ and $\texttt{Property } f \, s$ gives the set of couples of the property whose name is s, representing the values of the instances of the property attributed to the item referenced in the first element of a couple. If $s = (s_0, [s_1, \cdots, s_m])$, **Property** gives the set of couples of the property obtained by the composition of the properties $s_0, \cdots, s_n$. This is the case of structured values of a property s_0 that has (i.e. is described by) another property s_1 and so on. If f is sub or super, Property gives also the couples respectively of the sub or super-properties of s.

If the inverse switch is present, the inverse function must be used in place of the direct one: while the argument x represents the subject of the property specified by property, on the contrary x represents the object of the property specified by property inverse.

Formally replace Q with $Q' = \{(v, r) \mid (r, v) \in Q\}$ in the previous rule.

2.3 Textual Syntax and Semantics of MathQL-1 Query Results

The textual representations of query results belong to the grammatical production <mqr_set> whose semantics is described by four (infix) evaluating relations: $\Rightarrow_a$, $\Rightarrow_g$, $\Rightarrow_r$ and $\Rightarrow_s$.

$\Rightarrow_a$: <mqr_attr> $\to$ (String $\times T_0$) $\to$ Boole evaluates an attribute with a multiple string value. $\Rightarrow_g$: <mqr_group> $\to T_1 \to$ Boole evaluates a set of attributes with a their values. $\Rightarrow_r$: <mqr_res> $\to T_3 \to$ Boole evaluates resource with its sets of attributes. $\Rightarrow_s$: <mqr_set> $\to T_4 \to$ Boole evaluates a set of resources with their sets of attributes.

Note that a multiple string value can be empty. In fact, in an RDF data model a property can be optionally used, even if it is always declared in an RDF schema. Also note that the sets of attributes are always inhabited.

```
<mqr_attr>   ::= <vvar> [ "=" <string_list> ]?
<mqr_group>  ::= "{" <mqr_attr> [ ";" <mqr_attr> ]* "}"
<groups>     ::= <mqr_group> [ "," <mqr_group> ]*
<mqr_res>    ::= <string> [ "attr" <group> ]?
<mqr_set>    ::= [ <mqr_res> [ ";" <mqr_res> ]* ]?
```

Formally the evaluation works as follows:

$$\frac{i : \text{<vvar>} \quad q_1 : \text{<string>} \ ... \ q_m : \text{<string>}}{i = q_1, ..., q_m \Rightarrow_a (i, \{\text{Unquote } q_1, ..., \text{Unquote } q_m\})} \qquad \frac{x_1 \Rightarrow_a a_1 \ ... \ x_m \Rightarrow_a a_m}{\{x_1; ...; x_m\} \Rightarrow_g \{a_1, ..., a_m\}}$$

$$\frac{q : \text{<string>} \quad x_1 \Rightarrow_g A_1 \quad \cdots \quad x_m \Rightarrow_g A_m}{q \text{ attr } x_1, \cdots, x_m \Rightarrow_r (\text{Unquote } q, \{A_1, \cdots, A_m\})} \qquad \frac{x_1 \Rightarrow_r r_1 \quad \cdots \quad x_m \Rightarrow_r r_m}{x_1; \cdots; x_m \Rightarrow_s \{r_1, \cdots, r_m\}}$$

3 Implementation and Testing

3.1 Implementation

In this section we will briefly discuss the implementation of a MathQL-1 querying engine in the context of the HELM project describing the functions and relations available from the HELM metadata schemas[10] as well as other issues concerning the underlying database management system and the implemented software. Currently HELM metadata provide the following information on the mathematical resources of the library.

---

[10] <http://www.cs.unibo.it/helm/schemas/schema-h>,
   <http://www.cs.unibo.it/helm/schemas/schema-hth>.

- The standard Dublin Core metadata properties[11].
- The list of objects the object depends on. This information is available through a relation named refObj.
- The list of objects depending on the object. This information is available through a relation named backPointer.
- An alias for the reference of the object, given by the function shortName.

HELM is testing PostgreSQL[12] DBMS and Galax XQuery engine[13] as the support for the querying engine. The querying engine is written in Caml[14] for an easy integration with the other software developed for the HELM project, and currently it consists of the following parts.

- The *interpreter*, implemented by D. Lordi [2] and now re-implemented by L. Natile, executes a query given in its Caml representation (as a suitable data structure) giving back a Caml representation of the query result.
- The *input/output utilities* interface the textual or XML representation of a query or of a query result with its internal Caml representation.
- The *query generator* is the interface between the interpreter and the HELM proof assistant which needs to search the library for various purposes like interactive or automated proof searching. In particular the proof assistant provides a command (or tactic) named *SearchPatternApply* which uses the generator to issue a query for the set of statements that can possibly refine the current goal.

What follows is the textual representation of the query issued by the generator when the HELM proof assistant *SearchPatternApply* command is applied to the goal $2 * m \leq 2 * n$ where m and n are natural numbers and 2 is actually the successor of the successor of 0.

The references appearing in this query denote the following HELM resources:

"cic:/Coq/Init/Peano/le.ind#1/1"	less or equal
"cic:/Coq/Init/Peano/mult.con"	multiplication
"cic:/Coq/Init/Datatypes/nat.ind#1/1/2"	successor
"cic:/Coq/Init/Datatypes/nat.ind#1/1/1"	zero

```
let $positions be {"MainConclusion", "InConclusion"} in
let $universe be {"cic:/Coq/Init/Datatypes/nat.ind#1/1/1",
                  "cic:/Coq/Init/Peano/mult.con",
                  "cic:/Coq/Init/Datatypes/nat.ind#1/1/2",
                  "cic:/Coq/Init/Peano/le.ind#1/1"} in
  select @uri0 in
    select @uri in relation inverse "refObj"
      "cic:/Coq/Init/Peano/le.ind#1/1"
```

---

[11] <http://purl.org/dc/elements/1.1/>.

[12] <http://www.postgresql.org>.

[13] <http://db.bell-labs.com/galax/>.

[14] <http://caml.inria.fr>.

```
    attr $pos <- "position"
    where ex "MainConclusion" sub @uri.$pos
intersect
    select @uri in relation inverse "refObj"
    "cic:/Coq/Init/Peano/mult.con"
    attr $pos <- "position" where ex "InConclusion" sub @uri.$pos
intersect
    select @uri in relation inverse "refObj"
    "cic:/Coq/Init/Datatypes/nat.ind#1/1/2"
    attr $pos <- "position" where ex "InConclusion" sub @uri.$pos
intersect
    select @uri in relation inverse "refObj"
    "cic:/Coq/Init/Datatypes/nat.ind#1/1/1"
    attr $pos <- "position" where ex "InConclusion" sub @uri.$pos
where
    refof select @uri in relation "refObj" refof @uri0
           attr $pos <- "position"
           where ex $positions meet @uri.$pos
    sub $universe
```

The main operation in the query is the `select` `@uri0` and its `in` clause builds, through three `intersect`, the set of statements having each of these resources in their conclusions (in an appropriate position). The `where` clause is responsible for filtering out the statements whose conclusion refers to other resources than the ones above. Also note the exploitation of the `inverse` switch to invert the `refObj` relation. This query has some optimizations: the `let` constructions evaluate the sets `$positions` and `$universe` (this contains the resources that must appear in the statement), outside the `where` clause, which is evaluated many times during the execution of the `select`. Moreover the test:

```
$positions meet @uri.$pos
```

should be more efficient than the equivalent test:

```
"MainConclusion" sub @uri.$pos or "InConclusion" sub @uri.$pos
```

as this involves a larger number of operators.

3.2   Testing

The information currently available in the PostgreSQL database concerns 15244 HELM objects and comes from the scan of 416699 RDF statements stored in 80 Mb of disk space[15] while the database itself uses 382 Mb of disk space.

The following test evaluates the total processing time (including the query building time) of 165 queries (issued by the generator and processed by the Natile interpreter) like the one in the query example (see 3.1), with respect to the query complexity, which is proportional to the size of the `%universe` set.

---

[15] See [2] for a description of how the database is built from the RDF files.

The first table concerns PostgreSQL, the second table concerns Galax. This test shows that Natile engine handles complex queries more efficiently.

size	issued queries	time/size (mean)	time/size (variance)
1 to 2	59	0.25 sec.	0.23 sec.
3 to 9	106	0.03 sec.	0.02 sec.
1 to 9	165	0.11 sec.	0.17 sec.

size	issued queries	time/size (mean)	time/size (variance)
1 to 2	59	13.00 sec.	13.53 sec.
3 to 9	106	0.33 sec.	0.23 sec.
1 to 9	165	4.86 sec.	10.12 sec.

References

1. Asperti A., Padovani L., Sacerdoti Coen C., Guidi F., Schena I.: Mathematical Knowledge Management in HELM. In Electronic Proc. of MKM 2001.
   <http://www.emis.de/proceedings/MKM2001/>
2. Lordi D.: Sperimentazione e Sviluppo di Strumenti per la gestione di metadati. M. Thesis in Computer Science, University of Bologna. Advisor: A. Asperti. 2002
3. Rayavarapu S.: W3C Query languages, 29 January 2001.
   <http://www1.coe.neu.edu/~srayavar/W3CQL/ql.html>
4. Lassila O. and others: Resource Description Framework (RDF) Model and Syntax Specification, W3C Recommendation 22 February 1999.
   <http://www.w3.org/TR/1999/REC-rdf-syntax-19990222/>
5. Brickley D. and others: RDF Vocabulary Description Language 1.0: RDF Schema, W3C Working Draft 30 April 2002. <http://www.w3.org/TR/rdf-schema/>
6. Schena I. *Towards a Semantic Web for Formal Mathematics*. Ph.D. Thesis in Computer Science, University of Bologna. Supervisor: A. Asperti, February 2002.
7. Berners-Lee T. and others: Uniform Resource Identifiers (URI): Generic Syntax (RFC 2396). <http://www.ietf.org/rfc/rfc2396.txt>
8. Winskel G.: The formal semantics of programming languages: an introduction. MIT Press Series in the Foundations of Computing. London: MIT Press, 1993
9. Bray T. and others: Extensible Markup Language (XML) 1.0 (Second Edition), W3C Recommendation 6 October 2000. <http://www.w3.org/TR/REC-xml/>
10. Chamberlin D. and others: XQuery 1.0: An XML Query Language, W3C Working Draft 16 August 2002. <http://www.w3.org/TR/xquery/>

Information Retrieval in MML

Grzegorz Bancerek[1] * and Piotr Rudnicki[2]**

[1] Institute of Computer Science, Białystok Technical University, Poland, and
Dept. of Information Engineering, Shinshu University, Nagano, Japan.
bancerek@mizar.org
[2] Dept. of Computing Science, University of Alberta, Edmonton, Canada
piotr@cs.ualberta.ca

Abstract. MIZAR, a proof-checking system, is used to build the MIZAR Mathematical Library (MML). This is a long term project aiming at building a comprehensive library of mathematical knowledge. We describe issues concerning information retrieval, i.e., searching, browsing and presentation of MML contents. A web-based tool providing such functionalities is being implemented by G. Bancerek. We hope that our observations are helpful when solving similar problems for other repositories of formalized mathematics.

1 Introduction

The MIZAR language is a language used for the practical formalization of mathematics close to mathematical vernacular used in publications. The continual development of the MIZAR system, since 1973, has resulted in a language, software for checking the correctness of texts written in the language, numerous utility programs, a centrally maintained library of mathematics, and an electronic hyper-linked journal, all available on the Internet.[3] Introductory information on MIZAR can be found in [5–7]. For the rest of this paper, we assume that the reader is at least superficially familiar with these basic texts.

The development of MML[3] has been the main activity of the MIZAR project since the late 1980's, as it has been believed within the project team that only a substantial experience may help in improving the system. MML is a collection of MIZAR articles. An article is originally presented as a text-file and contains theorems and definitions, at this moment—August 2002—there are 725 articles in MML, occupying 55096 kB, containing 31942 theorems and 6110 definitions.

MML is still minuscule from the viewpoint of covering substantial chunks of mathematical knowledge. However, the problem of information retrieval in MML became a burning issue a long time ago. This issue takes a number of faces:

- **Searching**: When proving, we frequently ask the question: *Is this fact present in* MML? Quite frequently we also face a more troublesome question: *Has this notion or something similar already been defined?*

---

* Partially supported by JSPS P00025.
** Partially supported by NSERC grant OGP9207.
[3] http://mizar.org

A. Asperti, B. Buchberger, J.H. Davenport (Eds.): MKM 2003, LNCS 2594, pp. 119–132, 2003.

– **Browsing**: In order to get acquainted with the contents of MML one has to read it, and this reading can be substantially aided by proper browsing tools—we do not have anything fancier than linked hypertext in mind.
– **Presenting**: The presentation of formal mathematical texts should probably be as close as possible to the source language in which they have been written—at least from the viewpoint of potential authors. However, there is a need to present search results in some other form. Also, when browsing, we face presentation problems as seeing just one piece of text at a time is unsatisfactory. The problems of presentation, to a large extent, fall into the domain of graphical user interfaces and we will not discuss them here.

Little is known to us about information retrieval in other proof-assistants. David Delahaye [2] discussed the problem for the Coq library. His approach is based on type isomorphisms in the spirit previously applied in search tools for program modules written in languages of the ML family. While Delahaye aimed at soundness of his searching techniques, his work does not provide searching tools for all constructs of Coq. Our approach is different: we need a practical tool that can assist an experienced MIZAR user in searching the ever growing MML. After some small experiments, we have resigned from basing our tools on database systems like SQL—they seem too general on one hand and too restrictive on the other for this stage of our project.

2   MIZAR **Articles and the Structure of MML**

A MIZAR article is written as a text file and consists of two parts: the *Environment Declaration* and the *Text Proper*. The *Environment Declaration* consists of *Directives* concerning imports from the data base. The *Text Proper* is a sequence of *Sections* consisting of a sequence of *Text Items*: theorems and definitions together with their proofs.

An article accepted into MML gets a unique identifier. Each article is processed into several forms distributed together with the checking software:

– The source text of the article (the `.miz` file) which may change when the library undergoes a revision.
– An *abstract* of the article which is a text file (the `.abs` file) containing its public information. The *Text Proper* of an article contains *public* and *non-public* information. The public information are statements of: theorems, definitions and schemes. Their justifications and all other items from *Text Proper* are non-public and they leave no trace in the abstracts. The items of public information are labeled and these labels are used for references from other articles.
– *Library (data base) files* which are used when importing items from the article to the local environment of another article. These files are not meant to be read by MIZAR authors.

Every article accepted to MML gets its presence on the web in the electronic *Journal of Formalized Mathematics*[3], JFM for short, which is organized into annual volumes and each article is presented in a number of forms:

- A short summary provided by the author(s) in English.
- The abstract in html format with hyperlinks from the point of use of a notion to its definitions.
- The full text of the article.
- A postscript rendition of the abstract obtained through a mechanical translation into English and typesetting using LATEX.

As MML is continually revised, the electronic *Journal* is updated and presents the up to date version of the articles. There is an associated paper publication called *Formalized Mathematics* ISSN 1426-2630, which publishes the articles in the state of their inclusion into MML.

3   html Based Browsing

The hyperlinked version of MIZAR abstracts in JFM provides rudimentary but quite satisfactory browsing facilities; satisfactory at least for MIZAR authors. Consider [10, FINSEQ_3] and the following theorem[4] about a finite sequence p

```
theorem                                          :: FINSEQ_3:27
    n in dom p iff 1 <= n & n <= len p;
```

The underlined items are linked to their definitions and thus clicking on dom shows us the definition of the domain of a relation [11, RELAT_1] (and thus of a function and thus of a finite sequence)

```
let R;
func dom R -> set means                          :: RELAT_1:def 4
    x in it iff ex y st [x,y] in R;
```

and clicking on len brings us the definition of the length of a finite sequence [1, FINSEQ_1]

```
let p;
redefine func Card p -> Nat means                :: FINSEQ_1:def 3
    Seg it = dom p;
    synonym len p;
```

This style of html browsing seems quite satisfactory.
(Please note that while in, set, and = are MML built-in notions, they have technical introductions provided in [3, HIDDEN].)
    JFM provides the presentation of MIZAR abstracts also in TEXed translations into English. Thus the last definition can be seen also as

---

[4] In fact, not much of a theorem as its proof is immediate

```
n in dom p iff 1 <= n & n <= len p
    proof n in Seg len p iff n in Seg len p; hence thesis by FINSEQ_1:3, def 3; end;
```

However, we will use this fact as our driving example.

> In the sequel p, q, r are finite sequences. ...
>
> Let us consider p. Then $\overline{\overline{p}}$ is a natural number and it can be characterized by the condition:
>
> (Def. 3)   Seg $\overline{\overline{p}}$ = dom p.
>
> We introduce len p as a synonym for $\overline{\overline{p}}$.

4   Text Based Searching

The reference information needed by MIZAR authors when writing a new article is provided by the .abs files which are conveniently located in one directory in the system distribution. For lack of better tools one can employ textual searches using the tools from the grep family; these can be performed from an editor[5]. However, besides some usefulness such searches have serious drawbacks, even for people thoroughly familiar with MML.

4.1   Searching for Theorems

Suppose that we are looking for a theorem which for a natural number relates its elementhood in the domain of a finite sequence with its relationship to the length of the sequence. In other words, our goal is to find theorems like FINSEQ_3:27 quoted above. A search invoked as `grep "dom.*len" *.abs` results in 59 hits. But, we are interested in cases where there is some logical connective between the functors. We try `implies` in this way `grep "dom.*implies.*len" *.abs` and we get 14 hits, none of which are even remotely close to what we are looking for. Maybe we tried the wrong connective, let us try `iff` instead. Executing `grep "dom.*iff.*len" *.abs` seems to give us what we wanted

```
finseq_3.abs: n in dom p iff 1 <= n & n <= len p;
finseq_3.abs: n in dom p iff n - 1 is Nat & len p - n is Nat;
scmfsa_7.abs:    insloc k in dom Load p iff k < len p;
```

Does it? We have assumed a specific order of the functors while equivalence is symmetric so we try `grep "(dom.*iff.*len)|(len.*iff.*dom)" *.abs` which results in 10 hits. We may look through the results but on the other hand we may further narrow our search. The relationship we are looking for should involve the predicate `in` relative to `dom` and the predicate `<=` relative to `len`. So we try

```
    grep "(in.*dom.*iff.*<=.*len)|(<=.*len.*iff.*in.*dom)" *.abs
```

which results in

```
finseq_3.abs: n in dom p iff 1 <= n & n <= len p;
goboard2.abs:  holds 1 <= n & n+1 <= len f iff n in dom f & n+1 in dom f;
goboard2.abs:  holds 1 <= n & n+2 <= len f iff n in dom f & n+1 in dom f & n+2 in dom f;
msualg_8.abs:    1 <= n & n < len p iff n in dom p & n+1 in dom p;
```

---

[5] Josef Urban (http://mizar.org/people) has prepared a MIZAR mode for emacs with rudimentary browsing, this tool will finally incorporate semantics based retrieval which we discuss later.

This is somewhat better than the last time, but is it everything that is relevant to our interests? Hard to say.

When using grep we search single lines of text and not entire syntactic units of MIZAR. This is unsatisfactory. Also, since the output is just single lines from .abs files, in order to find the theorem names we have to look into corresponding articles; but this procedure is conveniently supported in editors like emacs. Of course, it is quite easy to build a tool that would return entire items from .abs files while still performing only a textual search (such tools were available in the past but never caught on). grep has been used for a long time in searching .abs files and in the hands of someone familiar with MML it is a powerful tool. However, such searches frequently return a large number of hits as asking precise questions is troublesome, since the same meaning can be carried out in MIZAR in a variety of ways.

4.2 Searching for Definitions

Applying textual searches when searching MML for a notion is more troublesome than for theorems. What shall we search for? Our first guess is may be to look for a symbol used to express such a notion. But here we can only guess the symbol with which the notion was expressed. For instance, we might guess that the predicate we look for has been expressed with the symbol <=. Taking into account that this symbol could have occurred as a synonym or an antonym, we inquire

```
grep "(pred|synonym|antonym).*(<=|>=)" *.abs
```

for which we get 35 hits. Notation in MIZAR is heavily overloaded and this creates a problem as it is not clear how to narrow our search.

Certainly, we may attempt searching for a notion in terms of symbols that could have been used in the definiens. This, as in the case of searching for theorems, is unsatisfactory, as we are never sure about how to form our "query".

5 Semantics Based Retrieval

The design and implementation of semantics based retrieval in MML is still under development but an experimental version is available at http://megrez.mizar.org. An exhaustive presentation of this searching technique would require a sizable and quite technical description. Therefore, we resort to giving the key ideas illustrated by some examples.

5.1 Format, Pattern, Constructor

MML is based on built-in notions of set theory: set, in, and st =, which are technically introduced in [3, HIDDEN] but given their meaning in [8, TARSKI]. MIZAR offers a number of definitional facilities. In most cases, a definition defines a new constructor used later in syntactic constructions, and gives its syntax and meaning. However, a definition may only redefine some aspects of a previously defined

Table 1. Constructors and notations

Constructor	Syntactic construction	Constructor kind/code	Notation kind/code	Vocabulary symbol tag
Aggregate	Structural term	`aggr`	`aggrnot`	G
Attribute	Adjective	`attr`	`attrnot`	V
Functor	Term	`func`	`funcnot`	O
Mode	Type	`mode`	`modenot`	M
Predicate	Atomic formula	`pred`	`prednot`	R
Selector	Structure selector	`sel`	`selnot`	U
Structure	Structure type	`struct`	`structnot`	G

constructor, and either introduce a new constructor or not (we will not discuss the issue any further here as it would require a lengthy technical exposition).

The syntactic *format* of a constructor specifies the symbol of the constructor and the place and number of arguments. The symbol can be placed in prefix, infix or postfix position and take numerous arguments. The format of a constructor together with the information about the types of arguments is called a *pattern* of the constructor. The formats are used for parsing and the patterns for identifying constructors. A constructor may be represented by different patterns as synonyms and antonyms are allowed. Available kinds of constructors are presented in Table 1 (the right hand side of this table will be discussed later).

As an example, let us consider the following definition from [9, ORDERS_1]

```
let A be RelStr; let a1,a2 be Element of the carrier of A;
pred a1 <= a2 means                                :: ORDERS_1:def 9
   [a1,a2] in the InternalRel of A;
synonym a2 >= a1;
```

which defines a predicate constructor. The format of this constructor is a symbol in the infix position with (syntactically) two arguments: the left and the right argument. The symbols used in this format were defined in vocabulary HIDDEN as: R<= and R>=, where R indicates a predicate symbol.

The pattern of this constructor provides the information about the types of its arguments. It turns out that the constructor really has three arguments: two explicit ones of type `Element of the carrier of A` and one hidden argument, a `RelStr`, that is recoverable from the types of the visible arguments. Of course, the names of arguments in definitions are irrelevant.

The same constructor format is used in the following definition in [12, YELLOW_2]

```
let J be set, L be RelStr, f, g be Function of J, the carrier of L;
pred f <= g means                                  :: YELLOW_2:def 1
   for j being set st j in J
      ex a, b being Element of L st a = f.j & b = g.j & a <= b;
synonym g >= f;
```

but the constructor pattern is different: there are two visible arguments but now we have two hidden arguments. A pattern of a constructor together with the constructor is also called a *notation*. In different environments the same pattern may identify different constructors and thus lead to different notations. Each of the above two definitions introduces one constructor and two patterns for the constructor, and thus we have two notations for each contractor. Note that besides the "main" notation using symbol `<=` we also introduce a synonymic notation using symbol `>=` but we still have only one constructor in each case.

Constructors play a central role in the semantic based information retrieval as the meaning is attached to the constructor and not to the format or pattern.

5.2   Resources

In contrast to the text based searching where the resource we searched through was unstructured text, now we introduce a variety of searchable resources. These resources mirror the information stored in internal (not meant to be read by humans) MML files and in other associated software tools. The complete list of defined resource kinds is presented in Table 2. However, in the rest of this paper we will essentially skip issues concerning schemes, registrations, symbols, formats and FMKeywords[6] . We use the term *items* in correspondence to the library items of MML that are stored in internal data base files and can be referenced from MIZAR articles.

In the sequel, we will write the names of resources (indeed, resource classes) capitalized in this Font as we do in Table 2. Resource kinds, resource codes and examples of the query language will be written in the typewriter `font`. All elements of a resource have unique names.

- A name of an article is an identifier.
- An element of a resource from the Items class is associated with an article. The article name is a part of the element name which has this structure

   Article-name : Resource-code Number

   For example, `FINSEQ_1:func 3` is the name of the 3rd functor defined in the named article (see Appendix A). We have a *Constructor-name* when the *Resource-code* is one of the constructor codes from Table 1.

Please note that every resource code can serve as a unique resource kind but the converse is not true: some kinds of resources correspond to groups of resource codes. Resource codes and resource kinds are used in the query language in a variety of roles.

---

[6] MIZAR abstracts are mechanically translated into English, typeset using L^AT_EX and publish in *Formalized Mathematics*. In this typesetting, the symbols and names that correspond to MIZAR notions are called FMKeywords; they are carefully chosen to be close to everyday mathematical practice. We expect that using FMKeywords for searching may be very helpful for inexperienced or casual readers of MML.

Table 2. Resources

Resource	Resource kind	Resource code
Items	item	
Constructors	constr	
Individual constructors	*in Table 1*	*in Table 1*
Notations	notat	
Individual notations	*in Table 1*	*in Table 1*
Statements	stat	
Theorem(s)	th	th
Definitional theorem(s)	def	def
Definiens(es)	dfs	dfs
Scheme(s)	sch	sch
Theorems *or* Definitional theorems	thdef	
Registrations	reg	
Existential registration(s)	exreg	exreg
Conditional registration(s)	condreg	condreg
Term adjective registration(s)	funcreg	funcreg
Articles	article	
Symbols	symbol	
Formats	format	
FMKeywords	keyword	

5.3 The Query Language

The query language is based on several basic notions: resource, element of a resource, lists and operations on lists. For simplicity, we will sometimes use the term "resource" where to be precise we should have written "an element of a resource." There are numerous operations on resource elements and numerous ways of creating lists. An operation on a resource returns a list of resources. An operation on a resource applied to a list of resources results in a list of resources obtained by applying the operation to each element on the list. There are some operations that are applicable only to lists. In this section we give only a brief review of some aspects of the query language.

Primitive queries, primitive lists. A primitive query returns a list of (elements of) resources.

Article query: `article Article-name` returns the list of Items that reference constructors introduced in the article *Article-name*.

Constructor query: `Constructor-name` returns the list of Items that reference the constructor named *Constructor-name*.

Enumerated list:

    { *Resource-or-item-name,..., Resource-or-item-name* }

Whenever *Resource-or-item-name* is just an identifier we need to indicate which resource is meant. For example, `{ SUBSET_1 }` is erroneous as it is ambiguous: it can be a name of an article or a symbol and in order to remove the ambiguity one has to use a qualifier, and thus `{ article SUBSET_1 }` returns the list containing one article. When all elements on the list are names of articles we can use global qualification as in `article { tarski, xboole_0 }`.

Global-list: `list of Item-kind` returns all Items of the given kind. For instance `list of pred` returns the list of all 625 predicate constructors.

Restricted global list:

> `list of Item-kind from ( Article-name | Article-list )`

returns all resources of the given kind from the given article(s). In

```
list of constr from article { tarski, xboole_0 }
list of constr from ( list of article )
list of constr
```

the first query returns the list of all constructors from the listed articles while the second and the third are equivalent and return the list of all constructors. In the above query, the *Article-list* can be a result of a query resulting in a list of articles.

Operations. There is a number of basic operations for each resource kind. These basic operations reflect the MIZAR inner representation of resources of the given kind while the representation corresponds to the syntax of the MIZAR language used to introduce an element of such a resource. Therefore, the names of operations are derived from terms used in MIZAR grammar. An operation applied to a resource returns a resource or a list of resources. All operations are syntactically applicable to all resources but a specific operation may return a non empty answer only to a specific resource. Almost all operations have inverse operations.

Let us consider an example of resource Notations. Each notation is expressed in a format and thus we have operation on a notation also called `format`, each notation denotes a constructor and thus operation `constructor`. Similarly, the operation `synonym` returns synonyms of a given notation (non empty) and the operation `antonym` returns antonyms of the notation but this list can be empty.

Since the semantics of MIZAR constructs is directly associated with constructors, by far the most interesting and useful operations concern constructors. There are numerous operations allowing us to inquire about the constituents of a single constructor: its notation, its definition, its redefinitions, its origin, etc. The following two operations are frequently used in queries:

`ref` : Items $\rightarrow$ Constructors* returns the list of constructors that are referenced from the given item.

`occur` : Constructors $\rightarrow$ Items* returns the list of items in which the given constructor occurs.

The latter operation is a default operation on a constructor.

An interesting subclass of operations is used to inquire about types of arguments or the type of result for items other than statements. In the current version of the system the type of an argument is not a resource but we can inquire about the constructors occurring in types of arguments with the operation loci ref : Items → Constructors*. This operation is similar to the operation ref described above but is restricted to the types of arguments (a formal argument is called a *locus*).

Applying operations. (Both prefix and postfix notations are supported.) The query `ABIAN:attr 1 occur` is an operation on a single constructor returning 74 items (equivalent to a constructor query). `ABIAN:attr 1 ref` is an operation on a single constructor returning just one constructor.

Every operation on a resource is applicable to a list of resources and results in a list.

> *Resource-list* | *Resource-operation*
> *Resource-list* & *Resource-operation*

The operation is applied to each resource on the list, first obtaining a list of lists and then in the first case returning their union and in the second the intersection of the lists. Please note that `ABIAN:attr 1 | ref` is a list operation as in this context `ABIAN:attr 1` is treated as a constructor query and not the name of a single constructor.

One category of operations on lists are filters. Every resource kind can be used as a filter and `grep` is used for textual filtering.

> *List-of-items* | *Resource-kind*
> *List-of-items* | grep *Grep-pattern*

The following query first computes the list of Items where both the listed constructors occur (& occur) and then filters theorems from the list.

> `{ XBOOLE_0:func 1, XBOOLE_0:func 3 } & occur | th`

The following two queries are equivalent and filter all theorems from the series YELLOW of articles and return 1018 theorems for further processing.

> `list of th | grep YELLOW`
> `list of item | grep YELLOW | th`

Compound queries are built according to the following syntax

> *Query* (and | or | butnot) *Query*

and their meaning should be clear from the example in Section 5.5.

Group queries are built according to the following syntax

> (atleast | atmost | exactly)
>   ((*Constructor-item* { , *Constructor-item* }) |
>   * (*Query*))

For instance

> `atleast ( XBOOLE_0:func 1, XBOOLE_0:func 3 )`

returns 16 definitions and theorems where both these constructors occur together (XBOOLE_0:func 1 is the empty set and XBOOLE_0:func 3 is set intersection).

5.4 Browsing and Reviewing Resources

The query `list of pred from article ORDERS_1` results in the following "clickable" list for further browsing

ORDERS_1:pred 1
ORDERS_1:pred 2
ORDERS_1:pred 3

Clicking on the first line gives

ORDERS_1:pred 1   component rank: 2,
 [Registrations], [Constructors], [Universe]
 [Search (fam ORDERS_1:pred 1) , Search (ORDERS_1:pred 1)],
 definition, definiens, notation, [Neighbourhood], MXA, ABS
 definition
   let A1 be RelStr;
   let A2 be Element of the carrier of A1;
   let A3 be Element of the carrier of A1;
   pred A2 <= A3;
 end;

Further details about this constructor are available for browsing

component rank Component rank of an item is 1 + the maximum of component ranks of constructors appearing in the item.

Registrations List of cluster Registrations needed by this constructor.

Constructors List of Constructors occurring in this item, i.e., the answer to `ref ORDERS_1:pred 1`.

Universe Shows the types of all terms, including arguments of attributes and types, gradually expanded to the primitive type of set.

Search (fam ORDERS_1:pred 1) List of Items in which this constructor or some of its redefined variants occurs, i.e., the answer to `fam ORDERS_1:pred 1`, 418 items.

Search (ORDERS_1:pred 1) List of Items in which this constructor occurs, i.e., the answer to `occur ORDERS_1:pred 1`, 256 items.

definition The definitional theorem behind this definition, ORDERS_1:def 9, i.e., the answer to `definition ORDERS_1:pred 1`.

definiens The definiens of this definition, used when proving by definitional expansion, ORDERS_1:dfs 6, i.e., the answer to `definiens ORDERS_1:pred 1`.

notation The list of introduced Notations: ORDERS_1:prednot 1 and its synonym ORDERS_1:prednot 2, i.e., the answer to `notation ORDERS_1:pred 1`.

Neighbourhood List of Constructors which occur together with this constructor in other items.

MXA The extended abstract of article ORDERS_1 (see Appendix A).

ABS The abstract of article ORDERS_1.

5.5   An Example of a Search for a Theorem

Let us try to perform a search with the same goal as in Section 4.1. In our first
attempt, we look for items that use the required constructors: RELAT_1:func 1
for dom and FINSEQ_1:func 3 for len and of course they must use FinSequence,
that is also FINSEQ_1:attr 1. The query

```
FINSEQ_1:attr 1 and RELAT_1:func 1 and FINSEQ_1:func 3
```

results in 23 items. For continuing in a sensible way some knowledge of MML is
required. The constructor for dom was redefined a number of times such that we
should probably allow for any of its derivatives, that is, for the entire family of
doms. Unfortunately, the query

```
FINSEQ_1:attr 1 and fam RELAT_1:func 1 and FINSEQ_1:func 3
```

results in 86 hits, much more than 23 but it could have been expected. We try to
narrow our search by also requiring the presence of the following constructors:
HIDDEN:pred 1 for in, which is needed to relate to dom and ARYTM:pred 2 for <=,
needed to relate to len

```
FINSEQ_1:attr 1 and fam RELAT_1:func 1 and FINSEQ_1:func 3
and HIDDEN:pred 2 and ARYTM:pred 1
```

and we are left with 14 items, some of them are not theorems, so we filter

```
( FINSEQ_1:attr 1 and fam RELAT_1:func 1 and FINSEQ_1:func 3
    and HIDDEN:pred 2 and ARYTM:pred 1 ) | th
```

They are presented to us in a "clickable" form such that a cursory examination
of some of them allows us to narrow our query by indicating which constructors
are not desirable, and thus we arrive at

```
( FINSEQ_1:attr 1 and fam RELAT_1:func 1 and FINSEQ_1:func 3
    and HIDDEN:pred 2 and ARYTM:pred 1 butnot FUNCT_1:func 1 ) | th
```

which results in 5 theorems, one more than with the textual search, namely
TOPREAL7:th 5.

5.6   An Example of a Search for a Notion

Let us try to find a notion related to "ordering" of functions, like the predicate
[12, YELLOW_2:pred 1] (see p. 6). We may reasonably suspect that such a predicate
is expressed with the predicate symbol <= or >=. The query

```
symbol {<=, >=} | notation
```

returns 18 predicate notations. But the predicate we look for should involve
the notion of a function and we first find the constructors behind the obtained
notations

```
symbol {<=, >=} | notation | constructor
```

which gives us 11 predicate constructors. The constructor we are after should
involve the notion of a function and thus FUNCT_1:attr 1, therefore we query

```
symbol {<=, >=} | notation | constructor and FUNCT_1:attr 1
```

and in return we obtain only 1 constructor YELLOW_2:pred 1. Unfortunately, the
story is not so simple as we have missed something. If we look through the

originally obtained notations then we will find notation FUNCT_4:prednot 2 which uses symbol <= and needs arguments that are functions. However, there is no new constructor behind this notation as it really denotes predicate constructor for the subset among sets and thus our last query missed it. The query that we should have used is only a bit more complex

```
symbol { <= , >= } | notation
                       where [ loci ref and { FUNCT_1:attr 1 } ]
```

6   Conclusions

The work on the semantic based retrieval in MML has started only recently. The importance of sophisticated tools for browsing and searching has become apparent in some sub-projects within MIZAR which involve many authors and many articles. The current, still experimental implementation turned out to be very helpful as we can now query MML in a way impossible with textual searches. At the moment, this retrieval tool is geared toward advanced users of MIZAR and plays a crucial role in creating EMM—Encyclopedia of Mathematics in MIZAR (see [4, XBOOLE_0, XBOOLE_1]). Articles in EMM have monographical character. The "raw" material from the original articles is *semi-automatically* extracted into encyclopedic articles with the substantial assistance of the presented retrieval machinery.

References

1. Grzegorz Bancerek and Krzysztof Hryniewiecki. Segments of natural numbers and finite sequences. *Formalized Mathematics*, 1(**1**):107–114, 1990.
2. David Delahaye. Information Retrieval in a Coq Proof Library using Type Isomorphisms. In T. Coquand *et al.*, editors, *TYPES*, volume 1956 of *LNCS*, pages 131–147. Springer, 2000.
3. Library Committee of the Association of Mizar Users. Mizar Built-in Notions. http://mizar.org/JFM/Axiomatics/hidden.html.
4. Library Committee of the Association of Mizar Users. Boolean Properties of Sets. http://mizar.org/JFM/EMM.
5. MIZAR *Manuals*. http://mizar.org/project/bibliography.html.
6. P. Rudnicki, Ch. Schwarzweller and A. Trybulec. Commutative Algebra in the Mizar System. *Journal of Symbolic Computation*, **32**:143–169, 2001.
7. Piotr Rudnicki and Andrzej Trybulec. On equivalents of well-foundedness. *Journal of Automated Reasoning*, 23(3-4):197–234, 1999.
8. Andrzej Trybulec. Tarski Grothendieck set theory. *Formalized Mathematics*, 1(**1**):9–11, 1990.
9. Wojciech A. Trybulec. Partially ordered sets. *Formalized Mathematics*, 1(**2**):313–319, 1990.
10. Wojciech A. Trybulec. Non-contiguous substrings and one-to-one finite sequences. *Formalized Mathematics*, 1(**3**):569–573, 1990.
11. Edmund Woronowicz. Relations and their basic properties. *Formalized Mathematics*, 1(**1**):73–83, 1990.
12. Mariusz Żynel and Czesław Byliński. Properties of relational structures, posets, lattices and maps. *Formalized Mathematics*, 6(**1**):123–130, 1997.

A   Mizar Extended Abstracts

The primary source of information about the contents of MML is the MIZAR
abstracts. They provide all the information necessary for referring to the MML
contents. Here is how the definition of the length of a finite sequence appears in
the abstract file (p denotes a FinSequence):

```
definition let p;
 redefine func Card p -> Nat means
:: FINSEQ_1:def 3
 Seg it = dom p;
 synonym len p;
end;
```

(Please note that the hyperlinked version of this definition was presented in
Section 3). Unfortunately, this plain version of MIZAR abstract does not provide
any information about the resources occurring in the corresponding library item.
To rectify this situation, Grzegorz Bancerek in his implementation generates the
so called *extended abstracts* in which the occurring resources are displayed in a
form of a comment. The above definition is presented in the extended abstract
as follows:

```
definition let p;
 redefine
:: FINSEQ_1:funcnot 3 => FINSEQ_1:func 3
:: FINSEQ_1:def 3
:: Constructors:
::     ARYTM:func 1, ARYTM:func 2, FINSEQ_1:attr 1, FINSEQ_1:func 2,
::     FINSEQ_1:func 3, FUNCT_1:attr 1, HIDDEN:mode 1, HIDDEN:pred 1,
::     RELAT_1:attr 1, RELAT_1:func 1, SUBSET_1:mode 2
func Card p -> Nat means
 Seg it = dom p;
:: FINSEQ_1:funcnot 4 => FINSEQ_1:func 3
 synonym len p;
end;
```

This definition redefines the functor CARD_1:func 1 which we can learn by
inquiring origin FINSEQ_1:func 3 (it is not clear that this information should be
shown directly in the extended abstract, the contents of extended abstracts is
still experimented with).

The above definition introduces one new functor FINSEQ_1:func 3 and two
notations for the functor: FINSEQ_1:funcnot 3 and FINSEQ_1:funcnot 4. These
names are only used for searching, the name FINSEQ_1:def 3 is used in MIZAR
texts to refer to the definitional theorem of this definition.

The constructors listed in the extended abstract of FINSEQ_1:def 3 are the
constructors that are obtained as a result of the query ref FINSEQ_1:def 3.

An Expert System for the Flexible Processing of XML–Based Mathematical Knowledge in a PROLOG–Environment

Bernd D. Heumesser[1], Dietmar A. Seipel[2], and Ulrich Güntzer[1]

[1] University of Tübingen, Wilhelm–Schickard Institute for Computer Science
Sand 13, D – 72076 Tübingen, Germany
{heumesser,guentzer}@informatik.uni-tuebingen.de
[2] University of Würzburg, Department of Computer Science
Am Hubland, D – 97074 Würzburg, Germany
seipel@informatik.uni-wuerzburg.de

Abstract. In this paper, we describe techniques for querying and transforming XML–based mathematical knowledge. The XML–documents are transformed into an equivalent PROLOG–structure called *field notation*, which serves as our *Document Object Model* (DOM).

Based on the field notation we provide a powerful and flexible *query language* in a PROLOG–based logic programming environment enabling *intelligent reasoning* about the data. It also offers a method which allows for elegantly encoding transfomations on XML–documents, using a powerful *substitution* mechanism.

We are applying these techniques in an *expert system* for the *classification* and the *retrieval* of ordinary differential equations. The rule–based approach allows to provide a *query* and *transformation language*, which can deal with different kinds of XML–based mathematical documents, such as documents in MATHML and in OPENMATH.

1 Introduction

XML–technology [12] changed the document formats in almost all application domains. The use of XML has led to a separation of presentation and content in many areas. Mathematical knowledge is more and more frequently represented in an XML–based document format. In fact, now the documents can not only contain the mathematical notation for presentation purposes, but can also contain the intended meaning of the described mathematical structure. Representing and storing mathematical achievements and knowledge in a self–describing, extensible and open manner, facilitates Web information systems and simplifies world–wide cooperation among mathematicians.

This knowledge is spread over the Web and it can only be handled if there is a possibility to operate on and to query this information based on the underlying mathematical structure. In the mathematical community there are efforts to make mathematical knowledge more easily usable via the Web. For instance the MOWGLI–project (cf. [1]) works on a development of a technological infrastructure for the creation and the maintenance of a virtual distributed, hypertextual

A. Asperti, B. Buchberger, J.H. Davenport (Eds.): MKM 2003, LNCS 2594, pp. 133–146, 2003.
© Springer-Verlag Berlin Heidelberg 2003

library of mathematical knowledge based on a content description of the information.

Using logic programming techniques for dealing with mathematical knowledge is an approach that has also been used by Dalmas, Gaëtano and Huchet [5], who present a deductive database MFD2 for mathematical formulas. The information stored in the database can be queried by using a special unification algorithm which takes care of the conditions associated with the formulas. The MBASE system of Kohlhase and Franke [8] provides access to different theorem proving systems based on systems using logic programming components. These two systems put the focus on using deduction for theorem proving.

We introduce a PROLOG–based environment for handling XML–based mathematical knowledge. We use two aspects of PROLOG: On one hand we apply PROLOG–like a procedural programming language to implement a basic library using, e.g., its capabilites to deal with terms and its unification mechanism. On the other hand we use its rule–based features and its inference–engine for deductions based on these libraries. The system is accessed by the HTTP–Protocol and can retrieve documents stored on the local file system or remote documents adressed by a URL.

The paper is organised as follows: First, we give a brief introduction to the modeling of mathematical knowledge in XML–languages. In Section 3, we present the PROLOG–Document Object Model (DOM) that is used in our expert system; it allows for handling XML–documents nicely in PROLOG. We introduce a query language called FNQUERY for accessing and for extracting parts of the documents. We also present a method which allows for elegantly encoding transfomations on XML–documents, using a powerful *substitution* mechanism. In Section 4, we show how our expert system can deal with different XML–languages, and we apply these techniques to the domain of ordinary differential equations. Finally, the conclusion gives a summary of our work and a short outlook.

2 XML–Based Mathematical Knowledge

XML enables markup languages – so–called applications of XML – to be defined for a given purpose or for an application domain. Since all of these languages share the same grammatical structure it is possible to have generic tools and software systems to deal with all XML–applications at least in a general way. Documents in each XML–application may be stored in an XML– (or an XML–enabled) database. Especially the use in the Web is supported by many tools enabling processing XML–documents for displaying in a standard Web–browser.

At this time the two major XML–languages for representing mathematical knowledge are MATHML and OPENMATH. They are *complementary* (cf. [4]). Therefore we are considering both languages in this paper.

2.1 The Content Markup of MATHML

The mathematical markup language MATHML [14] consists of two parts: One part is the presentation markup used for presentation purposes. In this part of

MATHML the user specifies how the content should be presented. This may differ from the mathematical structure, because for presentation it is not necessary to specify the exact mathematical structure. The second part of MATHML – the content markup – is a language in which mathematical structures may be defined in an exact way.

Since we are intereseted in the mathematical content of documents, we consider only the content markup of MATHML. But we have to deal with some presentation markup, too, because the two languages may be mixed, for instance if one uses variables with subscripts in the content markup. In MATHML, the set of available XML–tags is fixed. There are many predefined symbols, functions, and containers. However, it is possible to use other XML–languages like, e.g., OPENMATH inside a MATHML–structure.

2.2   The OPENMATH–Standard

OPENMATH [4] is another standard for representing mathematical knowledge. This language solely deals with the semantic meaning of the knowledge. OPENMATH uses an XML–language for exchanging mathematical objects; in this paper we refer to this representation.

In contrast to MATHML, OPENMATH defines only a small set of tags. They are just used as containers and the semantic information is given in their attributes. Hence the tag names carry only general information. On the other hand there is a concept to assign semantic information to the objects, namely *content dictionaries*. Content dictionaries hold the meaning of some mathematical symbols. These symbols are used by the OMS–tag via addressing the content dictionary with the cd–attribute and the symbol with the name–tag.

2.3   An Example for MATHML and for OPENMATH

Figure 1 shows the representation of the equation $y'(x) + y(x) \cdot \cos(x) = e^{2 \cdot x}$ in the content markup of MATHML on the right hand side and the content markup of OPENMATH on the left hand side; this equation might appear in a document in another XML–language, such as, e.g., OMDOC [9] or XHTML [13]. Observe that – as Figure 1 suggests – the general structure is almost the same in both languages.

In OPENMATH the semantic information is encoded in the values of the tags referencing a content dictionary and a symbol name, whereas the MATHML–tags are more readable for humans. E.g., the times–tag in MATHML is denoted in OPENMATH as a generic OMS–tag with two attributes for addressing the content dictionary and the symbol name.

3   XML and PROLOG

For the use inside of our expert system we have developed an *abstract data type* called *field notation* for representing XML–documents, which supports operations such as accessing children and attributes of XML–documents or assinging new

MathML	OpenMath

<pre><math>
  <apply>
    <eq/>
    <apply>
      <plus/>
      <apply>
        <diff/>
        <bvar>
          <ci>x</ci>
          <degree>
            <cn type="integer">1</cn>
          </degree>
        </bvar>
        <apply>
          <ci type="fn">y</ci>
          <ci>x</ci>
        </apply>
      </apply>
      <apply>
        <times/>
        <apply>
          <ci type="fn">y</ci>
          <ci>x</ci>
        </apply>
        <apply>
          <cos/>
          <ci>x</ci>
        </apply>
      </apply>
    </apply>
    <apply>
      <exp/>
      <apply>
        <times/>
        <cn type="integer">2</cn>
        <ci>x</ci>
      </apply>
    </apply>
  </apply>
</math></pre>
	<pre><OMOBJ>
	  <OMA>
	    <OMS cd="relation1" name="eq"/>
	    <OMA>
	      <OMS cd="arith1" name="plus"/>
	      <OMA>
	        <OMS cd="calculus1" name="diff"/>
	        <OMBIND>
	          <OMS cd="fns1" name="lambda"/>
	          <OMBVAR>
	            <OMV name="x"/>
	          </OMBVAR>
	          <OMA>
	            <OMV name="y"/>
	            <OMV name="x"/>
	          </OMA>
	        </OMBIND>
	      </OMA>
	      <OMA>
	        <OMS cd="arith1" name="times"/>
	        <OMA>
	          <OMV name="y"/>
	          <OMV name="x"/>
	        </OMA>
	        <OMA>
	          <OMS cd="transc1" name="cos"/>
	          <OMV name="x"/>
	        </OMA>
	      </OMA>
	    </OMA>
	    <OMA>
	      <OMS cd="transc1" name="exp"/>
	      <OMA>
	        <OMS cd="arith1" name="times"/>
	        <OMI>2</OMI>
	        <OMV name="x"/>
	      </OMA>
	    </OMA>
	  </OMA>
	</OMOBJ></pre>

Fig. 1. Representation of $y'(x) + y(x) \cdot \cos(x) = e^{2 \cdot x}$

values to them. It serves as our DOM for building a digital library of fine–grained mathematical objects.

Based on this field notation we provide a powerful and flexible *query language* in a PROLOG–based logic programming environment. This bridges the gap between XML–based digital libraries and deductive (inference–based) databases enabling intelligent reasoning about the data by providing full access to the PROLOG programming environment. We have also implemented a *transformation language* that is inspired by XSLT. It can for instance be used for transforming HTML–documents or XHTML–documents – which currently are available on the Web – to well–structured, semantically enriched XML–documents.

3.1   A PROLOG–DOM for XML

We use *association lists* for representing complex objects, cf. [11]. This data structure is familiar to LISP programmers. An association list consists of attribute/value–pairs of the form "$a_i : v_i$", where a_i is an attribute and "v_i" is the associated value. Thus, a complex object O, where $v_i = O.a_i$, $1 \leq i \leq n$, can be represented as an association list

$$O = [a_1 : v_1, \ldots, a_n : v_n].$$

Notice, that the values "v_i" themselves can be association lists (i.e., the definition is inductive) or atomic values. XML–objects can have attributes and sub–elements. Thus, an XML–object with the tag name "A" can be represented as a triple A : As : Es, where "As" is an association list for its attribute/value–pairs and "Es" represents its sub–elements. If there are no attributes, then the XML–object can alternatively be represented by A : Es instead of A : [ ] : Es. We call the notation

$$O = [a_1 : v_1 : w_1, \ldots, a_n : v_n : w_n] \tag{1}$$

for a list of XML–objects *field notation*. An XML–document can be represented by a list [A : As : Es] containing one element, or simply by that single element if no confusion arises. E.g., the XML–documents of Figure 1 are represented in field notation in Figure 2. Representing complex objects in field–notation rather than as ordinary PROLOG–atoms has got several advantages: Firstly, the sequence of attribute/value–pairs is arbitrary. Secondly, values can be accessed by attributes rather than by argument positions. Thirdly, the database schema can be changed at run time. Fourth, null values can be omitted, and new values can be added at run time. Fifth, semi–structured data, such as XML–data, can be represented and queried very elegantly; we will see this in the following.

Examples for FNQUERY. In the following we give some examples for the usage of the field notation and the language FNQUERY. A more detailed description can be found in [10].

For an object O of the form (1) in field notation we can select the list X = w_i of sub–elements using the statement "X := O^a_i", and we can select the list Y = v_i of attribute/value–pairs using the statement "Y := O@a_i". Here ":=" is an infix–predicate symbol with two arguments, which evaluates the terms "O^a_i" and "O@a_i", and assigns the result to the first argument "X" and "Y", respectively.

```
?- O = [ a:[d:1]:[ b:2, c:3], c:4 ],
   X := O^a,
   Y := O@a.

X = [b:2, c:3]
Y = [d:1]
```

MathML	OpenMath
| ```
math:[]:[
  apply:[]:[
    eq:[]:[],
    apply:[]:[
      plus:[]:[],
      apply:[]:[
        diff:[]:[],
        bvar:[]:[
          ci:[]:[x],
          degree:[]:[
            cn:[type:integer]:[1] ] ],
        apply:[]:[
          ci:[type:fn]:[y],
          ci:[]:[x] ] ],
      apply:[]:[
        times:[]:[],
        apply:[]:[
          ci:[type:fn]:[y],
          ci:[]:[x] ],
        apply:[]:[
          cos:[]:[],
          ci:[]:[x] ] ] ],
    apply:[]:[
      exp:[]:[],
      apply:[]:[
        times:[]:[],
        cn:[type:integer]:[2],
        ci:[]:[x] ] ] ] ]
``` | ```
OMOBJ:[]:[
  OMA:[]:[
    OMS:[cd:relation1, name:eq]:[],
    OMA:[]:[
      OMS:[cd:arith1, name:plus]:[],
      OMA:[]:[
        OMS:[cd:calculus1, name:diff]:[],
        OMBIND:[]:[
          OMS:[cd:fns1, name:lambda]:[],
          OMBVAR:[]:[
            OMV:[name:x]:[] ],
          OMA:[]:[
            OMV:[name:y]:[],
            OMV:[name:x]:[] ] ] ],
      OMA:[]:[
        OMS:[cd:arith1, name:times]:[],
        OMA:[]:[
          OMV:[name:y]:[],
          OMV:[name:x]:[] ],
        OMA:[]:[
          OMS:[cd:transc1, name:cos]:[],
          OMV:[name:x]:[] ] ] ],
    OMA:[]:[
      OMS:[cd:transc1, name:exp]:[],
      OMA:[]:[
        OMS:[cd:arith1, name:times]:[],
        OMI:[]:[2],
        OMV:[name:x]:[] ] ] ] ]
``` |

Fig. 2. Field Notation of the XML–Documents in Figure 1

This reminds of the evaluation of an arithmetic expression "X is 3 * (4+5)" in PROLOG where "is" is an infix–predicate symbol with two arguments, which evaluates the arithmetic term "3 * (4+5)" and assigns the result to the first argument "X".

It is possible to have *complex paths expressions* for selecting sub–elements:

```
?- O = [ a:[d:1]:[ b:2, c:3], c:4 ],
   X := O^a^b.
```

```
X = 2
```

In pure PROLOG the previous statement would look much more complicated.

A query selecting with multiple path expressions returns a list of objects, namely one object for each selector:

```
?- O = [ a:[ b:2, c:3], c:4 ],
   X := O^[a, c, a^b].
```

```
X = [[ b:2, c:3], 4, 2]
```

If a path expression contains variable symbols, then all ground instances of the path expression which are an allowed path in the queried object are generated (on backtracking):

```
?- O = [ a:[ b:2, c:3], c:4 ],
   X := O^a^Path.

X = 2, Path = b ;
X = 3, Path = c

?- O = [ a:[ b:2, c:3], c:4 ],
   2 := O^Path.

Path = a^b
```

Finally, using the operator "*", we can *assign new values* to attributes or elements as follows:

```
?- O = [ a:[d:1]:[ b:2, c:3], c:4 ],
   O_2 := O*[^a@d:5,^c:6],
   O_3 := O^a*[^b:1].

O_2 = [ a:[d:5]:[ b:2, c:3], c:6 ],
O_3 = [ b:1, c:3]
```

3.2   The Transformation Mechanism

To demonstrate the power of the transformation mechanism we show how easily it can be used to convert two different XML–languages. We have built a high level *substitution mechanism* on top of the substitution mechanism provided by PROLOG. These interleaved substitutions allow complex transformations on the documents to be encoded easily.

The following PROLOG–code converts the OPENMATH–document OM in field notation from Figure 1 and 2 to a MATHML–document MM in field notation:

```
Substitution_Symbols = [
   (eq:[]:[]) - ('OMS':[cd:relation1,name:eq]:[]),
   (plus:[]:[]) - ('OMS':[cd:arith1,name:plus]:[]),
   (times:[]:[]) - ('OMS':[cd:arith1,name:times]:[]),
   (cos:[]:[]) - ('OMS':[cd:transc1,name:cos]:[]),
   (exp:[]:[]) - ('OMS':[cd:transc1,name:exp]:[]),
   (diff:[]:[]) - ('OMS':[cd:calculus1,name:diff]:[]) ],
Substitution_Structure = [
   (apply:[]:[Op,
       bvar:[]:[ci:[]:[Var],
          degree:[]:[cn:[type:integer]:[1]]],X]) -
      ('OMA':[]:[Op,
          'OMBIND':[]:[
              'OMS':[cd:fns1,name:lambda]:[],
              'OMBVAR':[]:['OMV':[name:Var]:[]],X]]),
   (apply:[]:[ci:[type:fn]:[F]|Es]) -
      ('OMA':[]:['OMV':[name:F]:[]|Es]),
```

```
(ci:[]:[N]) - ('OMV':[name:N]:[]),
(cn:[type:integer]:[I]) - ('OMI':[]:[I]),
(apply:[]:Es) - ('OMA':[]:Es),
(math:[]:Es) - ('OMOBJ':[]:Es) ],
append( Substitution_Symbols, Substitution_Structure,
    Substitutions ),
fn_transform_elements(Substitutions,OM,MM).
```

As one can see, the main work that has to be done is to define the substitutions, which are then applied to the object in field notation within the predicate fn_transform_elements of our library. Each substitution consists of two parts separated by a minus sign, and it represents a context–sensitive rewriting rule: the right hand expression of a substitution is replaced by the expression on the left hand side.

There are two lists of substitutions: The first one contains the substitutions used for converting the symbols. There are no PROLOG–variables in this list because we just need to substitute the OPENMATH tags by the corresponding tags in MATHML. So we don't need the PROLOG–substitution mechanism here. In the second list of substitutions, Substitution_Structure, there are some variables and hence the PROLOG–substitution mechanism is used.

Now we will show how this works. Consider the following substitution:

```
(apply:[]:[ci:[type:fn]:[F]|Es]) -
    ('OMA':[]:['OMV':[name:F]:[]|Es])
```

The predicate fn_transform_elements tries to unify the parenthesized term behind the minus sign. This can only be done, if the content of the OMA–tag starts with a variable. This indicates that the variable with the name F is a function, which is applied to the other elements Es of the content. This construct is translated in MATHML by using an apply–tag and a ci–container for the variable. This variable is characterized by the type–attribute as a function. The remaining content Es of the OMA–tag is left unchanged.

Note that only the substitution for the derivation is a little bit complex – the others are rather straightforward.

Of course we can use the same technology for doing the backwards conversion from MATHML to OPENMATH. In [3] it is shown how this can be done. We just have to encode the substitutions in the way shown above. The substitution mechanism allows this to be done very easily in a PROLOG–based environment like our system.

A transformation from OPENMATH to MATHML is not always possible, since OPENMATH offers the possibility to define new symbols by using new content dictionaries. As the example above shows, a subset of both languages can be identified, which in fact can be transformed.

Another example for data extraction and transformation dealing with HTML–documents can be found in the appendix.

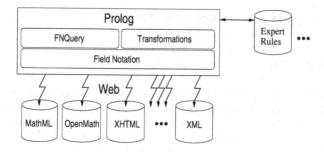

Fig. 3. Architecture of the Expert System

4 Applications in an Expert System

We are currently developing an expert system based on the techniques described in Section 3. It provides different kinds of operations on given documents: The digitally stored and edited mathematical knowledge can be accessed by applying a well–defined and structured access to the types and elements of the XML–documents. Transformations and queries encoded in this syntax are evaluated by the system and they can be mixed with PROLOG–predicates for performing reasoning tasks. I.e., it is possible to couple FNQUERY with the full programming facilities of PROLOG.

The rule–based approach allows to provide a *query* and *transformation language* which can deal with different kinds of XML–based mathematical documents, such as, e.g., documents in MATHML, OPENMATH, or yet to be defined XML–languages for special subject areas.

The following two subsections show how the techniques have been applied. The queries in Section 4.1 demonstrate that our system can handle MATHML– and OPENMATH–documents in a relatively uniform way. In Section 4.2 we apply the system for classifying ordinary differential equations; a more detailed description of this second application is given in [6].

4.1 Application to MATHML and to OPENMATH

We are in the process of extending FNQUERY to an abstract, easy–to–use query language which can be transformed to FNQUERY and then processed by the system to give the user the possibility to define his own queries. The queries generated by the system should cover both MATHML– and OPENMATH–documents. Because of the different properties of both languages, we have to identify a subset of corresponding structures and concepts which we would like to support.

Let `Object` be an XML–document – represented in field notation – containing some sub–elements in MATHML. Consider the following PROLOG–rule to extract differential equations from the document:

```
query_1(Object,Equations) :-
  findall( Equation_1,
    ( Equation_1 := Object^_^math,
```

```
        Equation_2 := Equation_1^apply,
        _ := Equation_2^eq,
        _ := Equation_2^_^diff ),
    Equations ).
```

It extracts a list – possibly with duplicates – of MATHML–objects which represent an equation containing the diff–operator. As one can see, this is encoded in just 4 lines of FNQUERY–predicates: The first matches only math–tags. The second line ensures that the following tag is an apply–tag. Then it is checked that the content of the apply–tag (directly) contains an eq–tag. At last it is proven that somewhere in this object there is a diff–tag.

Note that the above procedure only works for MATHML. But is is quite easy to transform it to deal with OPENMATH. We just have to use the OPENMATH terminology. This results in the following PROLOG–rule to extract all equation objects containing the diff–operator from an OPENMATH–document Object in field notation:

```
query_2(Object,Equations) :-
    findall( Equation_1,
        ( Equation_1 := Object^_^'OMOBJ',
          Equation_2 := Equation_1^_,
          [relation1,eq] :=
              Equation_1^['OMA'@'OMS'^cd,'OMA'@'OMS'^name],
          [calculus1,diff] :=
              Equation_2^['OMA'@'OMS'^cd,'OMA'@'OMS'^name]),
        Equations ).
```

The predicates look somewhat more complex, but this is only caused by the fact that we have to access the two attributes of the OMS–tag and to check both. Since PROLOG–names starting with capital letters are variables, we have to enclose tag names starting with a capital letter with single quotes.

The next step is to combine the predicates used in the above query to obtain predicates working for both languages. In addition, these predicates may easily be extended to deal with other XML–based mathematical markup languages. This is done like shown in the following rules:

```
math_object_contains(Object,diff) :-
    _ := Object^_^apply^diff.
math_object_contains(Object,diff) :-
    [calculus1,diff] :=
        Object^_^['OMA'@'OMS'^cd,'OMA'@'OMS'^name].
```

These two rules act as alternatives. The predicate succeeds if one of these two rules lead to success. If a set of these predicates is available, then they can be used to implement complex queries. Therefore it is easy to combine this rule–based approach with FNQUERY, the transformations mentioned above and the programming facilities of PROLOG within our expert system.

4.2   Application to Ordinary Differential Equations

We have applied these techniques to the *classification* and the *retrieval* of ordinary differential equations based on Kamke's rules. Kamke provides an ordered list of ordinary differential equations, cf. Section C of [7]. A user searching for a solution of a given equation has to transform the equation according to a set of rules. The form of the transformed equation determines rather precisely the place of the differential equation within Kamke's list. Thus the user has to browse only a very small list of possible candidates, where he might find his equation in the exact form or a special case of it.

Our expert system supports the search in this collection. We are in the process of defining criteria for the equations to shrink the set of candidates in the list which can match the given equation (in the sense given above) to a reasonable size; for more details see [6]. If this can be done in an effective way, then we can try to unify the given equation with the remaining candidates. This might be done by incorporating efficient *term–rewriting* methods into our system.

5   Conclusions

We access the semantic information encoded in XML–based mathematical knowledge for providing querying and transformation techniques. Embedded in the PROLOG environment, our rule–based system uses the field notation as its Document Object Model, which results in the natural and powerful query language FNQUERY, and in a mechanism for the transformation of documents based on substitutions.

This approach features complex reasoning and retrieving tasks which can be implemented in short time in a very powerful language. The rule–based approach allows for a rather generic implementation, which can deal with almost arbitrary XML–languages that might be used for representing mathematical knowledge. The application of our expert system to the collection of ordinary differential equations shows that this approach is feasible.

Other concepts for transforming and querying XML–documents are the languages XSLT [16] and XQUERY [17]. Both use the language XPATH [15] for addressing objects in an XML–document. Even though they are very powerful languages, our PROLOG–based approach has got some advantages, since the embedding of FNQUERY in PROLOG yields the full access to the potentials of using the PROLOG–predicates or self–defined predicates based on the PROLOG programming language. Hence, very complex computations and transformations are possible. A more detailed description of the power of FNQUERY and a comparison to other query or transformation languages dealing with XML will be the topic of future research. Our goal is to develop an abstract query language built on top of FNQUERY and to implement a graphical, Web–based user interface.

According to Berners–Lee et al. (cf. [2]), the goal of the *Semantic Web* is to add *a logic component for reasoning* about the Web. Therefore embedding techniques for dealing with XML in a logic programming language like PROLOG is a step in this direction.

References

1. A. Asperti, B. Wegner. MOWGLI – *A New Approach for the Content Description in Digital Documents.* Proc. of the 9th Intl. Conference on Electronic Resources and the Social Role of Libraries in the Future, Section 4, Volume 1, 2002.
2. T. Berners–Lee, J. Hendler, O. Lassila. *The Semantic Web.* Scientific American, May 2001.
3. D. Carlisle, J. Davenport, M. Dewar, N. Hur, W. Naylor. *Conversion between* MATHML *and* OPENMATH. Bath/NAG, 2001.
4. O. Caprotti, D. P. Carlisle, A. M. Cohen. *The* OPENMATH *Standard.* The OPENMATH Esprit Consortium, February 2000.
5. S. Dalmas, M. Gaëtano, C. Huchet. *A Deductive Database for Mathematical Formulas* In: J. Calmet, C. Limongelli (Eds.): Design and Implementation of Symbolic Computation Systems, Springer LNCS, 1996.
6. B. Heumesser, D. Seipel, R. Schimkat, U. Güntzer. *A Web-Information System for Retrieving and Reasoning about* XML–*Based Mathematical Knowledge* EICM'2002, `http://www-db.informatik.uni-tuebingen.de/forschung/talk_eic.shtml`.
7. E. Kamke. *Differentialgleichungen – Lösungsmethoden und Lösungen.* Akademische Verlagsgesellschaft, 8th Edition, 1967.
8. M. Kohlhase, A. Franke. MBASE: *Representing Knowledge and Context for the Integration of Mathematical Software Systems.* Journal of Symbolic Computation, Volume 32, Number 4, 2001.
9. M. Kohlhase. OMDOC: *An Open Markup Format for Mathematical Documents.* Technical Report, Department of Computer Science, Carnegie Mellon University, Pittsburgh, March 2002, `http://www.cs.cmu.edu/~kolhase`.
10. D. Seipel. *Processing* XML–*Documents in* PROLOG. Workshop on Logic Programming WLP'2002.
11. D. Suciu, S. Abiteboul, P. Bunemann. *Data on the Web – From Relations to Semi–Structured Data and* XML. Morgan Kaufmann, 2000.
12. *Extensible Markup Language (*XML*) 1.0,* World Wide Web Consortium, October 2000. `http://www.w3.org/TR/2000/REC-xml-20001006`.
13. XHTML *2.0,* World Wide Web Consortium, August 2002, `http://www.w3.org/TR/2002/WD-xhtml2-20020805/`.
14. *Mathematical Markup Language (*MATHML*) Version 2.0,* World Wide Web Consortium, February 2001, `http://www.w3.org/TR/MathML2/`.
15. XML *Path Language (*XPATH*),* World Wide Web Consortium, November 1999, `http://www.w3.org/TR/xpath`.
16. XSL *Tranformations (*XSLT*),* World Wide Web Consortium, August 2002, `http://www.w3.org/TR/xslt20/`.
17. XQUERY *1.0: An* XML *Query Language,* World Wide Web Consortium, August 2002, `http://www.w3.org/TR/xquery`.

Appendix

We show an example of extracting information from an HTML–document to show the flexibility of the substitution mechanism. Processing HTML–documents is a more difficult task than processing XML–documents. This example shows that our system can even deal with HTML.

Knowledge Extraction from the MathML–Tutorial

The following example has been taken from the MathML–tutorial at

http://www.dessci.com/support/tutorials/mathml/default.stm

which can be reached from http://www.w3.org/Math/. The goal was to extract the examples from the section on *Containers and Operators*. The HTML–document contains sequences – describing examples – of the following form:

```
<p>
  <b>Expression:</b>
    <p>
      <img SRC="misc/set-interval.gif" width="229" height="30">
        <p>
          <b>Markup:</b>
          <pre>
            <reln> <eq/>
              <set>
                <bvar>
                  <ci>x</ci>
                </bvar>
                <condition>
                  <reln> <leq/>
                    <cn>0</cn> <ci>x</ci>
                  </reln>
                </condition>
              </set>
              <interval closure='closed-open'>
                <cn>0</cn> <ci>&infty;</ci>
              </interval>
            </reln>
          </pre>
```

The SGML–parser of SWI–PROLOG parses the HTML–document into a PROLOG–atom. When the parser reaches the end of the document, then it automatically closes all open tags. This is important, since the HTML–tags <p> and <img> typically are not closed. Thus, the SGML–parser creates PROLOG–structures, where, e.g., the pre–element is nested within the img–element.

The following PROLOG–predicate extracts the image and the markup for all examples. Firstly, it selects a p–element "X" at any depth, such that "X" contains a b–element with the contents 'Expression:'. Secondly, the attribute list "Img" of the image is selected. Thirdly, the markup "Markup" is selected as the sub–element of "X" that is formatted by "pre". Observe that this sub–element occurs nested within the img–element, since this was not closed in HTML. Since this element was formatted by "pre", it is parsed into a PROLOG–atom "XML_Atom" by the SGML–parser of SWI–PROLOG. This PROLOG–atom can be parsed into field notation using our predicate xml_atom_to_fn_term.

```
fn_term_to_mathml_examples(FN_Term,Examples) :-
    findall( example:[expression:[img:Img:[]],markup:Markup],
        ( X := FN_Term^_^p,
          ['Expression:'] := X^b,
          Img := X^p@img,
          [XML_Atom] := X^p^img^p^pre,
          xml_atom_to_fn_term(XML_Atom,Markup) ),
        Examples ).
```

The following PROLOG–predicate transforms all elements with the tag "SRC" by replacing the tag by "src".

```
fn_mathml_transform_elements(FN_Term_1,FN_Term_2) :-
    Substitutions_1 = [
        (src:Source) - ('SRC':Source) ],
    fn_transform_elements( Substitutions_1,
        FN_Term_1, FN_Term_2 ).
```

Thus we obtain the following XML–element, where the reln–element is the one from the original HTML–document.

```
<example>
  <expression>
    <img src="misc/set-interval.gif" width="229" height="30"/>
  </expression>
  <markup>
    <reln> ... </reln>
  </markup>
</example>
```

Obviously, the original HTML–document could also be simplified by closing all p– and all img–tags. But it turns out that the processing of the modified HTML–document would not be easier for our system.

Towards Collaborative Content Management and Version Control for Structured Mathematical Knowledge

Michael Kohlhase[1] and Romeo Anghelache[2]

[1] Computer Science, Carnegie Mellon University, kohlhase+@cs.cmu.edu

[2] Albert Einstein Institute, Golm, Germany, romeo@psyx.org

Abstract. We propose an infrastructure for collaborative content management and version control for structured mathematical knowledge. This will enable multiple users to work jointly on mathematical theories with minimal interference.

We describe the API and the functionality needed to realize a CVS-like version control and distribution model. This architecture extends the CVS architecture in two ways, motivated by the specific needs of distributed management of structured mathematical knowledge on the Internet. On the one hand the one-level client/server model of CVS is generalized to a multi-level graph of client/server relations, and on the other hand the underlying change-detection tools take the math-specific structure of the data into account.

> Versioning is a can of worms.
> But what good is a can of worms if you never open it?
> *Norm Walsh on the* www-tag *mailing list, 11 Sep. 2002*

1 Introduction

In the last years we have seen the birth of a new research area: "Mathematical Knowledge Management" (MKM), which is concerned with representation formalisms for mathematical knowledge, such as MATHML [CIMP01], OPEN-MATH [CC98] or OMDOC [Koh00], mathematical content management systems [FK00,ABC+02,APCS01], as well as publication and education systems for mathematics. The perceived interest in the domain of general knowledge management tools applied to mathematics is that mathematics is a very well-structured and well-conceptualized subject. The main focus of the MKM techniques is to recover the content/semantics of mathematical knowledge and exploit it for the application of automated knowledge management techniques, with an emphasis on web-based and distributed access to the knowledge.

In this paper, we extend the focus of MKM techniques from the distributed *access* to mathematical knowledge to the *creation process* of mathematical knowledge, which is — for the most part — a distributed and collaborative process. After all, even if mathematicians often develop individual contributions

A. Asperti, B. Buchberger, J.H. Davenport (Eds.): MKM 2003, LNCS 2594, pp. 147–161, 2003.
© Springer-Verlag Berlin Heidelberg 2003

alone (e.g. in single-authored papers), the progress of a mathematical theory or sub-field involves a multitude of authors — communicating via meetings, messages and publications. Moreover, in contrast to the "knowledge access" scenario, where the mathematics is relatively static, the "knowledge creation" scenario involves managing the change of resources. We claim that MKM techniques have the potential of supporting this scenario as well, and that the "knowledge creation" scenario is potentially even more important for applications, as knowledge can only be accessed after it has been created. In fact, we expect the implementations of techniques like the ones presented in this paper, to play a similarly facilitating role in the development of open repositories of formal mathematical knowledge as the code management systems like the CVS system [CVS] have had for the creation of the wealth of open-source software we know today.

Currently, MKM systems either support simple monotonic addition of mathematical content or are specialized to particular applications, e.g. the Maya system [AHMS02] which is specialized to formal software engineering and verification. The "development graph" model for a management of theory change [Hut00] employed in this system uses a rich set of relations among theories to trace logical dependencies among mathematical objects and propagate/limit the effects of changes to the theories.

Our own MKM system MBASE [FK00,KF01] is currently a member of the first class, but it can communicate with the MAYA system via the joint interface language OMDOC [Koh00]. As an effect, MBASE/MAYA support theory management on the fragment of OMDOC that corresponds to the MAYA development graph. In fact, in [KF01] we have proposed to distinguish two kinds of MBASEs, different in their data changing policies.

- An archive MBASE which is epitomized by the Journal MBASE MJ in our scenario below, it archives unchanging mathematical knowledge and is referenced by many other MBASEs.
- A scratch-pad MBASE like the personal MBASEs MR and MR', that do not have any dependents and are primarily used for theory development.

To get a feeling of the requirements for the functionality addressed in this paper, let us take a look at a likely research communication scenario: We will first describe the communication pattern in a neutral way — say as it could have happened in the era of mathematics done with pen and paper (around 2001), and then model it using distributed MKM (about 2010).

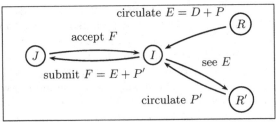

Fig. 1. Classical Research Cooperation

Classical, see Figure 1. Researcher R works on theory T together with his colleague R' at institute I. The theory T is a body of mathematics laid down in an article A published in journal J. Now, R extends theory T by a new definition D (say for a mathematical object O), proves a set P of theorems about O, and calls the resulting extended theory E. After that, R tells her colleague R' at I about D and P (say by circulating a memo in I), who gets interested and proves a set P' of useful properties of O. Together, R and R' put the theory E into final form F, and submit it to journal J. This accepts F and publishes it.

With MKM, see Figure 2. In 2005, the publisher of journal J has established an MBASE server MJ for J which now contains theory T. Furthermore, the institute has its own departmental MBASE MI and the researchers R and R' have the personal MBASEs MR and MR'. Now R develops the

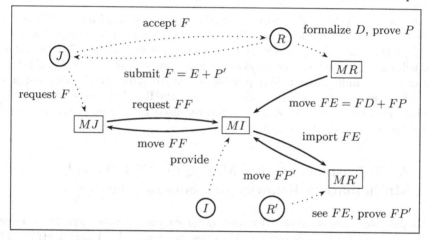

Fig. 2. Research Cooperation with distributed MKM

formalization FD of O, stores it in MR and formalizes the set P of theorems by formalizing them and formally proving them[1] (yielding FP in MR). Instead of sending around an internal note about D and P in I, R moves their formalizations FD and FP into the institute MBASE server MI, from where R' can import them into his personal MBASE MR'[2]. On this basis R' formally proves FP', and adds it to theory FE, yielding FF the formal version of theory F. Then R and R' submit F to journal J, who evaluates it (possibly via his own personal MBASE) it and finally accepts F. To publish F on MJ, it requests FF from MI, which moves it there.

---

[1] To do so, R may need to revise the initial version of D several times in order to be able to prove the desired theorems (reproving the already obtained results that depended on a previous version of D every time). This process is supported by MBASE/MAYA based on techniques presented in [AHMS02]

[2] Alternatively, R could leave FD and FP in MR and tell R' personally about them, allowing him to import them from MR into MR'; but this is a matter of institute policy, which we will not address here.

1.1   Contribution of This Paper

As we have seen in the scenario above, a strict division into archive and scratch-pad knowledge bases is unrealistic, since it does not reflect the current and anticipated nature of scientific communication and publication: Collaboration and theory change occurs at every level and should be supported by an infrastructure for collaborative content management and version control which enables multiple users to work jointly on mathematical theories with minimal interference.

We will develop a general architecture for a collaborative content management extending the CVS architecture and specialize it for mathematical knowledge by taking into account the structure of mathematical documents. For the second task we will build on both the work on structural diff/patch/merge utilities in XML, as well on the semantic management of change in the MAYA system [AHMS02].

Even though the work reported in this paper is motivated by the MBASE system, it is much more general, since it only depends on the communication format used by the system. The methods are not even specific to the OMDOC format, we will only assume that the knowledge base systems use a similar XML-based format for communication and provide a way to re-create the original interface documents. This would for instance cover the the HELM system [APCS01], which employs a lightweight infrastructure based mainly on XML documents and XSL(T) stylesheets for MKM.

2   An Infrastructure for Managing Distributed Mathematical Knowledge Cooperatively

The proposed infrastructure for collaborative theory management is largely based on the CVS (**C**oncurrent **V**ersions **S**ystem [CVS]) architecture. This system is widely used to support collaborative software development, since it combines software versioning with controlled concurrent access to the resources under CVS control. We will briefly review the basic notions of CVS, and describe our multi-level architecture with reference to it.

2.1   Cooperative Version Control in CVS

CVS is a server-based system for concurrent version control, used mainly for software development. The CVS server provides a so-called CVS repository R, which keeps a representation of all committed versions (called *revisions*) of the software together with logging information.

A CVS client C can then *check out* a *working copy* of the software and work on it. Let us for simplicity assume that C checks out the most recent revision in the repository, the so-called *head*. After completing the development task, C can *commit* the changes Δ to the repository, creating a new (current) revision in the repository. She will usually accompany the commit with a short description of the changes; this is also logged in the repository, eventually adding up to a changelog for the software development.

It is a distinguishing feature of CVS that the repository is not locked when a working copy is checked out. So another client C' can also have active working copies of the software and work on them. When C commits, the working copy of C' which was based on the (old) head, is no longer *up to date* with the repository. As a consequence, the changes C' has made to the software cannot be committed to the repository. C' can not simply check out a new working copy from R, since she would lose her work; therefore (upon C's request) CVS merges the changes Δ into C's working copy to keep it up to date with respect to the head of R. Now (after resolving any conflicts introduced by the merge) C' can commit her changes to the repository. Even though conflicts can occur in the merging operation, they are sufficiently infrequent in practice.

We have seen above that version control in CVS protocol is based on the computation, communication and management of differences (changes) to files. CVS uses the unix utilities

diff for determining the changes in a working copy to be committed to the repository

patch for updating old revisions

merge for merging changes into a working copy to keep it up to date with the repository.

To facilitate the functionality described above, the CVS server represents committed non-head revisions of files internally as reverse diffs from the head revision (which is stored explicitly). Thus the head revision can be served immediately, whereas older revisions can be computed by applying the respective reverse diffs. In this model, a version can be represented as a specific sequence of transformations (edit scripts).

2.2   A Multi-level Client/Server Architecture

CVS has a one-level client-server architecture, i.e. all the CVS clients can only communicate with a dedicated CVS server. In the distributed MKM settings like in Figure 2, we have a knowledge base MI that acts both as a repository for MR and MR' and as a client for MJ.

We will say that a knowledge base A is *downstream* from an knowledge base B, iff A is a CVS client of B or any knowledge base C that is downstream from B. The relation of being *upstream* is the converse relation to *downstream*. In Figure 2, MR, MR', and MI are downstream from MJ. Note that commit actions push information upstream and update actions pull information downstream.

A multi-level client-server architecture has inherent advantages: it can, for instance simulate CVS branching: In CVS a branch is used, if a set of clients want to make changes to software that are either too disruptive or too extensive for the usual update/commit cycle. In essence a branch acts as a virtual repository for the development and allows controlling revisions without disturbing the main development (the so-called *trunk*).

In our multi-level architecture, a branch in repository A for clients $S^1, \ldots, S^n$ can be simulated by creating a new knowledge base B downstream from A and upstream from the S^i. B is initialized by checking out a working copy from A, and the S^i can track their revisions in B and eventually commit the result to B. Closing the branch corresponds to deleting the knowledge base B and updating the S^i.

2.3   An Atomized Version Control Relation

The CVS protocol is based on the file system hierarchy for grouping and anchoring user interaction. For instance, update and commit commands issued without reference to a particular file will be applied to all registered files in the current directory.

The file system hierarchy is replaced with a document-centered (given by omgroup in the OMDOC representation) or semantic hierarchies (given by theories or development graphs). The notion of a file (or equivalently of an OMDOC document) is only a secondary concept — if present at all — in the conceptual hierarchy of mathematical knowledge management systems. In particular, the level of a file is not the lowest level of an object under version control. This role is taken up by the notion of a mathematical object represented by a top-level OMDOC element. As a consequence, the client/server relation is atomized to mathematical objects instead of files. We speak of the *version control relation* that relates working copies of mathematical objects with their repository instances. Of course this relation must be acyclic.

Just as a file system can contain working copies from multiple repositories, a knowledge base can contain objects that are working copies checked out different repositories, though for each mathematical object, the version control relation is a tree, i.e. every object has at most one server it can be committed to and updated from. Intuitively, a math object is — for the parent element — as a file for a directory; and files have attributes like creation time, modification time, permissions; so should the math objects, which can be stored in the Dublin Core metadata of OMDOC elements.

2.4   Interaction of Version Control with Distribution and Knowledge Base Consistency

In [KF01] we have identified four tasks necessary for distributing mathematical knowledge bases: caching, moving, changing, and deleting mathematical objects. Before we give them interpretations in our architecture, let us re-examine the assumptions we based the analysis on; they include (paraphrased):

A3 all mathematical elements have a unique *"defining"* realization in the network of knowledge bases.

A4 mathematical objects are never changed.

Assumption **A3** is directly related to distribution: every object has a unique description: a pair consisting of the URL of the knowledge base and the unique identifier of the object there. All other copies of the object are just cached copies of it.

Assumption **A4** was useful for distribution, since it makes caching and maintenance very simple. Relaxing **A4** — which is the task at hand in this paper

— has two aspects: How do we ensure consistency in situations where e.g. a definition or theorem that other mathematical objects depend on are changed in mathematically significant ways[3]. We will not deal with this problem here, since it is already studied in great detail in the development graph model [Hut00].

The question we will address in this paper is purely at a protocol level: it can largely be framed in terms of the interaction between **A3** and version control. We will study this with respect to the three distribution tasks identified in [KF01].

Caching Mathematical Objects: We used assumption **A4** to allow trivial caching. In the new architecture, we identify the caching relation to be the version control relation: to cache a copy of a mathematical object, it is simply checked out from the repository as a working copy. Note that objects that are working copies can never be defining instances of mathematical objects in our model. In the new model cache-consistency is a well-understood problem, since an object can always be updated from its repository. The ensuing conflicts can be resolved by the standard three-way merge methods described e.g. in [Lin01].

Moving Mathematical Objects: One of the most basic procedures is that of moving objects between knowledge bases, e.g. of the theory FF from MI to MJ after the submission described in our scenario. This action can be modeled by adding FF as a defining instance to MJ, deleting FF in MI, and checking out FF from MJ to MI, which acts as a CVS client for MJ for this object. Note that with this construction, we can only move mathematical objects upstream, which is the natural direction.

Deleting and Changing Mathematical Objects: Since we leave the question of maintaining knowledge base consistency to the development graph techniques which entail re-examining mathematical objects that depend on the changed ones, augmenting the "pull" technology of our CVS-like architecture with a "push" component seems advantageous. Note that mathematical objects are always upstream from ones that logically depend upon them. Therefore a knowledge base $\mathcal{M}$ keeps a record of all the upstream knowledge bases, so that these can be notified of any changes and trigger propagation of the change. Apart from notification of dependents this information can be used for optimizations like the following: Whenever $\mathcal{M}$ moves the defining instance of an object $\mathcal{O}$ to some knowledge base $\mathcal{M}''$, then it can send the new location of $\mathcal{O}$ to all upstream knowledge bases, asking them to update their reference objects and thus shielding itself from future requests to $\mathcal{O}$.

3 Computing Differences and Managing Change

In this section we will describe the computational utilities underlying our collaboration architecture. CVS uses the line-based `diff/patch/merge` utilities to compute differences between versions, update files, and merge differences into

---

[3] Of course changes like correcting typos or changing explanatory text are unproblematic from a consistency point of view.

modified working copies. In applications like ours, where we know more about the structure of the data, we can do better, and arrive at more compact, less intrusive edit scripts[4]. For instance, if we know that whitespace carries no meaning in a document format, two documents are considered equal, even if they differ (with respect to the distribution of whitespace characters) in every single line; as a consequence, the computed difference would be empty.

We will look at different document models and their impact on computing differences between documents in this section. Before we do this, let us briefly clarify what we mean by a document model by comparison to mathematical models. In mathematics, when we define a class of mathematical objects (e.g. vector spaces), we have to say which objects belong to this class, and when they are to be considered equal (e.g. vector spaces are equal, iff they are isomorphic). For document models, we do the same, only that the objects are documents. XML supports the first task by allowing us to specify a document type definition (DTD) or an XML Schema, which can be used for mechanical document validation, but leaves the second to be clarified in the (informal) format specifications.

Listing 1. An OMDOC definition.

```
<definition id="comm−def" for="comm">
  <CMP xml:lang="en">An operation <OMOBJ id="op"><OMV name="op"/></OMOBJ>
    is called commutative, iff
    <OMOBJ id="comm1">
      <OMA><OMS cd="relation1" name="eq"/>
        <OMA><OMV name="op"/><OMV name="X"/><OMV name="Y"/></OMA>
        <OMA><OMV name="op"/><OMV name="Y"/><OMV name="X"/></OMA>
      </OMA>
    </OMOBJ> for all <OMOBJ id="x"><OMV name="X"/></OMOBJ>
    and <OMOBJ id="y"><OMV name="Y"/></OMOBJ>.
  </CMP>
  <CMP xml:lang="de">
    Eine Operation <OMOBJ xref="op"/> heißt kommutativ, falls
    <OMOBJ xref="comm1"/> für alle <OMOBJ xref="x"/> und <OMOBJ xref="y">.
  </CMP>
</definition>
```

Of course, the stronger the equality modulo which differences are computed, the better the edit scripts become. The conceptual core of the MBASE data model is given by the OMDOC format [Koh00,OMD], which is also used as an interface representation for communication between MBASEs and their clients. We will base our discussion in this section concretely on the OMDOC document model, building up to it by discussing the underlying XML document model. We will discuss generalizations to other document formats for MKM in section 3.4.

Let us call two documents $\mathcal{M}$-equal, iff they are equal with respect to the document model $\mathcal{M}$, analogously we will call an algorithm an $\mathcal{M}$-diff algorithm, iff it computes differences modulo $\mathcal{M}$-equality. In the rest of this section, we will use the OMDOC element in Listing 1 as a running example.

---

[4] Compactness of edit scripts is important for storage and query efficiency in MKM systems, while minimal intrusiveness (patching does not disrupt document structure) is important for humans to track and understand changes.

3.1 Using the Tree Structure of XML Documents

As OMDOC is an XML application, we can make use of the generic tree structure of XML documents. For instance, XML specifies that the order of attribute declarations in XML elements is immaterial, double and signle quotes can be used interchangeably for strings, XML comments (`<!--...-->`) are ignored, and whitespace characters in the UniCode serialization is only meaningful in text nodes. As a consequence, the serialization in Listing 2 is XML-equal to the one in Listing 1, but not to the one in Listing 4.

Listing 2. An XML-equal serialization for Listing 1

```
<definition  for="comm"                       id="comm-def" >
  ...
  <CMP xml:lang='de'> <!-- note the unabbreviated empty element -->
  Eine Operation <OMOBJ xref="op"></OMOBJ> heißt kommutativ, falls
  <OMOBJ xref='comm1'/> für alle <OMOBJ xref="x"/> und <OMOBJ xref='y'>.
  </CMP>
</definition>
```

There is a large body of work on using the XML tree structure to compute differences of XML documents modulo XML-equality (see e.g. [WDC02]). The algorithms (see [CRGMW96] for an introduction) compute partial tree matchings[5] and express these as so-called "edit scripts" that add and delete XML elements and attributes in the source tree to arrive at the target tree. The work has been mainly concerned with finding algorithms for optimal (least-cost) edit scripts and complexity issues. Formats like XUpdate [LM00] (see Listing 3 for an example) use XPATH [Cla99] expressions to identify the elements the instructions act upon.

The central problem of finding corresponding nodes in trees critically depends on the notion of tree-similarity employed. If the document is strongly keyed (e.g. all elements have unique ID attributes, which cannot be changed by the user[6] or the knowledge management system employs some node numbering system like the one proposed in [CTZZ01]), then the key structure gives a very natural notion of node correspondence, and differencing becomes relatively simple. For the un-keyed case, only the notion of structural isomorphism and of ordered and un-ordered trees has been considered e.g. in [CRGMW96].

Listing 3. An XUpdate edit script (partly) updating Listing 1 to Listing 4

```
<xu:modifications xmlns:xu="http://www.xmldb.org/xupdate">
  <xu:variable name="c" select="definition/CMP[0]/OMOBJ[@id='comm1']"/>
  <xu:remove select="definition/CMP[0]/OMOBJ[@id='comm1']/@xref"/>
  <xu:append select="definition/CMP[0]/OMOBJ[@id='comm1']" child="1">
    <xu:value-of select="$c"/>
  </xu:append>
  <xu:remove select="definition/CMP[0]/OMOBJ[@xref='comm1']/*"/>
  <xu:update select="definition/CMP[0]/OMOBJ[@xref='comm1']/@xref">
    <xu:value-of select="'comm1'"/>
  </xu:update>
</xu:modifications>
```

---

[5] Which nodes correspond to each other modulo a given notion of tree similarity?

[6] The action of changing keys in the data, can lead to un-intuitive and computationally sub-optimal edit scripts, but does not compromise the method per se.

3.2 The OMDOC Document Model

Let us now take a look at how the OMDOC document model can be used for more *semantic* differencing (OMDOC-diff[7]).

Listing 4. An OMDOC-equal representation for Listings 1 and 2

```
<definition id="comm−def" for="comm">
 <CMP xml:lang="de">Eine Operation <OMOBJ xref="op"/> heißt kommutativ, falls
  <OMOBJ id="comm1">
   <OMA><OMS cd="relation1" name="eq"/>
    <OMA><OMV name="op"/><OMV name="X"/><OMV name="Y"/></OMA>
    <OMA><OMV name="op"/><OMV name="Y"/><OMV name="X"/></OMA>
   </OMA>
  </OMOBJ> für alle <OMOBJ xref="x"/> und <OMOBJ xref="y">.
 </CMP>
 <CMP xml:lang="en">An operation <OMOBJ id="op"><OMV name="op"/></OMOBJ>
  is called commutative, iff <OMOBJ xref="comm1"/> for all
  <OMOBJ id="x"><OMV name="X"/></OMOBJ> and
  <OMOBJ id="y"><OMV name="Y"/></OMOBJ>.
 </CMP>
</definition>
```

The OMDOC document model extends the XML document model in various ways. For instance[8], the order of CMP children of an omtext element does not matter, and the distribution of whitespace is irrelevant even in text nodes. More generally, as OMDOC documents have both formal and informal aspects, they can contain *data-set-based* as well as *document-structured* information. At one extreme an OMDOC document contains a formalization of a mathematical theory, as a reference for automated theorem proving systems. There, logical dependencies play a much greater role than the order of serialization in mathematical objects. We call such documents *data set based* and specify the value DataSet in the Type element of the OMDOC metadata for such documents. On the other extreme we have human-oriented presentations of mathematical knowledge, e.g. for educational purposes, where didactic considerations determine the order of presentation. We call such documents *document-structured* and specify this by the value Text. Note that since OMDOC allows to specify Dublin Core metadata [WKLW99] at many levels, document-structured and data set based parts can interleave in the same document, allowing OMDOC-diff algorithms to take this into account.

Moreover OMDOC uses a variant of OPENMATH objects [CC98] that can be represented as directed acyclic graphs (DAGs; using ID/IDREF links) rather than regular trees: an empty element with an xref attribute is OMDOC-equal to the element that carries the corresponding id attribute. As a consequence, the representations in Listings 1 and 2 are OMDOC-equal to the one in Listing 4, and an OMDOC-diff algorithm must generate the empty edit script between

---

[7] Note that we are *not* proposing to use mathematical equality here, which would make the formula $X + Y = Y + X$ (the OMOBJ with id="comm1" in Listing 4 instantiated with addition for op) mathematicallly equal to the trivial condition $X + Y = X + Y$, obtained by exchaning the right hand side $Y + X$ of the equality by $X + Y$, which is mathematically equal (but not OMDOC-equal).

[8] As an introduction to the OMDOC format is beyond the scope of this paper, we will assume a basic knowledge of [Koh00] and the material at [OMD].

all three, while an XML-`diff` algorithm should generate an extension of the
XUpdate script in Listing 3.

In particular, the process of exploding the DAG to a tree representation or
sharing a tree to a DAG should not result in a difference computation. The
same applies to the OMDoc representation of proofs, where an additional level
of structure sharing is possible. A case where the underlying structure of the
data is not tree-like, that is, not based on structure-sharing, is the development
graph itself, which can even be cyclic. Here, first steps for defining a correspondence relation and for determining changes have been taken in [AHMS00] and
implemented in the MAYA system.

3.3 Challenges for OMDoc-`diff` Algorithms

As we have shown, taking advantage of OMDoc-equality in computing differences leads to more concise edit-scripts, which is essential in an environment
where document processing applications manipulate mathematical content by
acting on internal data structures and generate target documents from these. In
such situations, it is impossible to predict which of the possibly many OMDoc-
equal representations will be generated. Since in a CVS-like collaborative protocol any `diff` can lead to a conflict that will require human intervention for
resolution, the availability of such algorithms will be crucial for adoption.

Of course extending XML-equality to OMDoc-equality in computing differences breaks the underlying assumptions of the algorithms described in section 3.1. For instance, the DAG-nature of OMDoc documents requires the differencing algorithms to (virtually) expand the objects to tree form while processing
them[9].

It seems that techniques from [BKTT02] can be used to get around the
obvious computational difficulties involved in differencing modulo equality.
[BKTT02] trivialize the tree matching problem by assuming that all tree representations are "strongly keyed", employing a generalized notion of data base
keys to determine element correspondence in XML documents. They claim that
sensible data formats are almost always strongly keyed up to data in XML
text nodes. We have not verified this for OMDoc yet, but for instance even
though `CMP` nodes do not have `ID` attributes, they are keyed, since they have
`xml:lang` attributes, which must be unique among their siblings. However, `CMP`
content however is not keyed, since it is generic text data (which is trivially un-
keyed) mixed with representations of mathematical object represented as content
MATHML or OPENMATH objects (this also caused some addressing problems
in the XUpdate script in Listing 3). Note that in OMDoc documents managed
by MKM systems (as opposed to directly written by hand), the OMDoc `mid`
attributes can be used for keying, alleviating the higher computational costs of
the un-ordered algorithms somewhat.

Obviously, we need a combination of the XML tree-based un-keyed algorithms with key-sensitive techniques for our application; such algorithms have
been requested, but to the author's knowledge not been reported on so far.

---

[9] In the file system metaphor, this would correspond to following symbolic links

3.4 Modular $\mathcal{M}$-diff Algorithms

Given that most of the OMDoc document model is rather standard (DAGs vs. trees, sets vs. lists of children, etc.), it is appealing to develop general $\mathcal{M}$-diff algorithms, where the notion of $\mathcal{M}$-equality is specified externally, e.g. by extending the document schema to a full document model.

Note that XML Schema so far only specifies full document models (i.e. including equality) for so-called *data types* (e.g. "100" and "1.0E2" are equal as members of the data type float). Thus we could define the notion of XML-Schema-diff, which would take these into account, but this is only marginally relevant for our problem here, since it only concerns the leaves of the trees we are dealing with.

Listing 5. Specifying Order in XML Schema using xs:appinfo

```
<xs:complexType name="omtextType">
  ...
  <xs:sequence>
    <xs:annotation><xs:appinfo><mdiff:unordered/></xs:appinfo></xs:annotation>
    <xs:element name="CMP" type="inCMPtype" maxOccurs="unbounded"/>
    <xs:element name="FMP" type="FMP" minOccurs="0" maxOccurs="unbounded"/>
  </xs:sequence>
  ...
</xs:complexType>
```

A more promising avenue seems to be to make use of the xs:appinfo[10] element to specify document models for complex types in XML Schemata — as opposed to just content models for validation. Based on the examination of the OMDoc document model in section 3.2, it seems plausible to assume that we could go a long way by specifying

Document order e.g. by an element mdiff:unordered in Listing 5, and
Link semantics e.g. as in Listing 6 where we specify that the xref attribute of an OpenMath object means that it represents a copy of the object that carries the corresponding id attribute.

Listing 6. Specifying DAG attributes in XML Schema using xs:appinfo

```
<xs:attributeGroup name="DAG.attrib">
  <xs:attribute name="xref" type="xs:anyURI" use="optional">
    <xs:annotation><xs:appinfo><mdiff:dag−source/></xs:appinfo></xs:annotation>
  <xs:attribute>
  <xs:attribute name="xref" type="xs:ID" use="optional">
    <xs:annotation><xs:appinfo><mdiff:dag−target/></xs:appinfo></xs:annotation>
  <xs:attribute>
</xs:attributeGroup>
```

So, if an XML document (fragment) is an instance of a schema that contains document model specifications like the ones in Listings 5 and 6, then a modular diff algorithm can read the schema and customize — multiple times during the parsing process if necessary — the comparison criteria used by the algorithm.

---

[10] The xs:appinfo is introduced in XML Schema expressly for such purposes.

4   Conclusion

We have laid down first ideas for a collaborative version control model for MKM systems, based on the OMDoc format. We have sketched the overall architecture, and determined some of the requirements for OMDoc-`diff` algorithms that come from the respective structural invariants of the data in MKM systems.

We have seen that the architecture can be kept quite close to that of the well-known CVS system[11], and interacts well with the requirements for distribution identified in [KF01], which is encouraging from an implementation point of view. In particular, we are currently experimenting with the idea to annotate all information necessary for a CVS-like file-based formalism in the `metadata` elements of mathematical objects. We could for instance use the existing Dublin Core `Date` and `Identifier` element for timestamping, and keeping version information. Further information, such as pointers to the repository in working copy objects can be kept in the `metadata/extradata` element provided by OMDoc expressly for this purpose. We will experiment with a HELM [APCS01]-like setup based on OMDoc files on web-servers and implement merging by server-side XSL(T) processing.

The main item for further research is an OMDoc-`diff` algorithm as described in section 3.2. In the literature on version management in XML, we often hear the argument that difference-computation is not needed in practice, since documents are generated by XML structure editors, but this only moves the burden from an independent postprocess (implement once) to a module in every editor. Moreover, this would penalize authors for using general XML editors, since they could only incorporate XML-`diff` algorithms. Finally, the actual editing process employed by the user may not correspond to the optimal edit script.

Given a good difference computation algorithm, merging can be obtained by relatively simple extensions, especially since our CVS-like architecture allows the usage of the so-called three-way merge (see [Lin01]), where two revisions are compared with respect to a known base revision, from which they have been created. Here, edit scripts for the changes from the base can be computed for both revisions. These can be analyzed and combined to a joint edit script which updates the base revision to the merged revision. [Man01,Lin01] present algorithms for three-way merge of XML documents and there are even commercial implementations (e.g. the one described in [LF02]). Since the merge operation only depends on the edit scripts which act on the generic XML structure, and not on the particular structure of the OMDoc format, we can use these algorithms and implementations off the shelf.

Acknowledgements. The first author was supported by a Heisenberg stipend of the German Research Agency (DFG). The authors would like to thank Serge Autexier, Andreas Franke, Dieter Hutter, and Bernd Krieg-Brückner for stimulating discussions in the early stages of this work.

---

[11] Actually, [BKTT02] propose a repository organization that is not `diff`-based, which would be interesting to experiment with, but integrating it into a *collaborative* version control environment is not trivial

References

[ABC+02]   Stuart Allen, Mark Bickford, Robert Constable, Richard Eaton, Christoph Kreitz, and Lori Lorigo. FDL: A prototype formal digital library – description and draft reference manual. Technical report, Computer Science, Cornell, 2002. http://www.cs.cornell.edu/Info/Projects/NuPrl/html/FDLProject/02cucs-fd\%l.pdf.

[AHMS00]   Serge Autexier, Dieter Hutter, Heiko Mantel, and Axel Schairer. Towards an evolutionary formal software-development using CASL. In C. Choppy and D. Bert, editors, Proceedings Workshop on Algebraic Development Techniques, WADT-99. Springer, LNCS 1827, 2000.

[AHMS02]   Serge Autexier, Dieter Hutter, Till Mossakowski, and Axel Schairer. The development graph manager MAYA (system description). In Hélène Kirchner, editor, Proceedings of 9th International Conference on Algebraic Methodology And Software Technology (AMAST'02). Springer Verlag, 2002.

[APCS01]   Andrea Asperti, Luca Padovani, Claudio Sacerdoti Coen, and Irene Schena. HELM and the semantic math-web. In Paul B. Jackson Richard. J. Boulton, editor, Theorem Proving in Higher Order Logics: TPHOLs'01, volume 2152 of LNCS, pages 59–74. Springer, 2001.

[BKTT02]   Peter Buneman, Sanjeev Khanna, Keishi Tajima, and Wang Chiew Tan. Archiving scientific data. In ACM SIGMOD International Conference on Management of Data (SIGMOD), 2002.

[CC98]   Olga Caprotti and Arjeh M. Cohen. Draft of the Open Math standard. The Open Math Society, http://www.openmath.org, 1998.

[CIMP01]   David Carlisle, Patrick Ion, Robert Miner, and Nico Poppelier. Mathematical Markup Language (MathML) version 2.0. W3C recommendation, World Wide Web Consortium, 2001. Available at http://www.w3.org/TR/MathML2.

[Cla99]   XML path language (XPath) Version 1.0. W3C recommendation, The World Wide Web Consortium, 1999. available at http://www.w3.org/TR/xpath.

[CRGMW96]   Sudarshan S. Chawathe, Anand Rajaraman, Hector Garcia-Molina, and Jennifer Widom. Change detection in hierarchically structured information. In ACM SIGMOD International Conference on Management of Data (SIGMOD), pages 493–504, 1996.

[CTZZ01]   Shu-Yao Chien, Vassilis J. Tsotras, Carlo Zaniolo, and Donghui Zhang. Storing and querying multiversion XML documents using durable node numbers. In Proc. of The 2nd International Conference on Web Information Systems Engineering (WISE), 2001.

[CVS]   Concurrent versions system: The open standard for version control. Web site at http://www.cvshome.org.

[FK00]   Andreas Franke and Michael Kohlhase. System description: MBASE, an open mathematical knowledge base. In David McAllester, editor, Automated Deduction – CADE-17, number 1831 in LNAI, pages 455–459. Springer Verlag, 2000.

[Hut00]   Dieter Hutter. Management of change in structured verification. In Proceedings Automated Software Engineering (ASE-2000). IEEE Press, 2000.

[KF01]   Michael Kohlhase and Andreas Franke. MBase: Representing knowl-
              edge and context for the integration of mathematical software systems.
              *Journal of Symbolic Computation; Special Issue on the Integration of
              Computer algebra and Deduction Systems*, 32(4):365–402, 2001.

[Koh00]   Michael Kohlhase. OMDoc: An open markup format for mathematical
              documents. Seki Report SR-00-02, Fachbereich Informatik, Universität
              des Saarlandes, 2000. http://www.mathweb.org/omdoc.

[LF02]   Robin La Fontaine. Merging XML files: A new approach providing
              intelligent merge of xml data sets. In *XML Europe 2002 - Conference
              Proceedings*, 2002.

[Lin01]   Tancred Lindholm. A 3-way merging algorithm for synchronizing or-
              dered trees – the 3DM merging and differencing tool for XML. Master's
              thesis, Helsinki University of Technology, 2001.
              http://www.cs.hut.fi/~ctl/3dm/.

[LM00]   Andreas Laux and Lars Martin. XUpdate - XML update language.
              Working Draft of the XML:DB Initiative, 2000.
              http://www.xmldb.org/xupdate/xupdate-wd.html.

[Man01]   Gerald W. Manger. A generic algorithm for merging SGML/XML-
              instances. In *XML Europe 2001 - Conference Proceedings*, 2001.

[OMD]   The OMDoc repository. web page at http://www.mathweb.org/omdoc.

[WDC02]   Yuan Wang, David J. DeWitt, and Jin-Yi Cai. X-Diff: An effective
              change detection algorithm for XML documents, 2002. Submitted,
              http://www.cs.wisc.edu/~yuanwang/xdiff.html.

[WKLW99]   S. Weibel, J. Kunze, C. Lagoze, and M. Wolf. Dublin Core Metadata
              Element Set, Version 1.1: Reference Description. DCMI Recommenda-
              tion, 1999. http://dublincore.org/documents/1999/07/02/dces/.

On the Integrity of a Repository of Formalized Mathematics

Piotr Rudnicki[1]* and Andrzej Trybulec[2]**

[1] Dept. of Computing Science, University of Alberta, Edmonton, Canada
piotr@cs.ualberta.ca
[2] Institute of Informatics, University in Białystok, Białystok, Poland
trybulec@math.uwb.edu.pl

Abstract. MIZAR, a proof-checking system, is used to build the MIZAR Mathematical Library (MML). This is a long term project aiming at building a comprehensive library of mathematical knowledge. The language and the checking software evolve, and the evolution is driven by the growing library. We discuss the issues of maintaining integrity of an electronic repository of formal mathematics, based on our experience with MML.

1 Introduction

The MIZAR language is a language used for the practical formalization of mathematics. The main goal for the design of MIZAR was to create a formal system close to the mathematical vernacular used in publications with the requirement that the language be simple enough to enable computerized processing, in particular mechanical verification of correctness. The continual development of MIZAR has resulted in a language, software for checking the correctness of texts written in it, numerous utility programs, a centrally maintained library of mathematics, and an electronic hyper-linked journal, all of which are available on the Internet[1]. The logic of MIZAR is classical and the proofs are written in a natural deduction style. Definitions allow for the introduction of constructors for types, terms, adjectives, and atomic formulae. Introductory information on MIZAR can be found in [15,20,21,22,27]. For the rest of this paper, we assume that the reader is at least superficially familiar with these basic texts.

Around 1988, a decision was made within the MIZAR project to focus on developing the mathematical library—MIZAR Mathematical Library (MML[1])—based on the Tarski-Grothendieck set theory. We would like to stress the distinction between MIZAR as a formal system of general applicability (which as such has little in common with any set theory) and the specific application of MIZAR in developing MML which is based entirely on a set theory (see [24, TARSKI]). In building MML, the MIZAR language, the assumed logic, and the chosen set

---

* Partially supported by NSERC grant OGP9207.
** Partially supported by FP5 grant HPRN-CT-2000-00102.

[1] http://mizar.org

A. Asperti, B. Buchberger, J.H. Davenport (Eds.): MKM 2003, LNCS 2594, pp. 162–174, 2003.

theory provide an environment in which all further mathematics is developed. This development is *definitional*: new mathematical objects can be defined only after supplying a model for them in the already available theories.

MML is a collection of MIZAR articles. It would be difficult to define what makes a MIZAR article worthwhile for inclusion in MML; the final, typically quite liberal, decision is with the Library Committee. An article is stored as a text-file and contains theorems and definitions, at this moment—August 2002, there are 725 articles in MML, occupying 55096 kB, containing 31942 theorems and 6110 definitions. Theorems and definitions in MML are technical terms and correspond to mathematical facts, no matter how simple or how deep.

Building and maintaining such a body of formalized mathematics as MML forces a number of decisions and poses a number of challenges: choosing foundations, soundness, organization of the repository, searching facilities, presentation and browsing, maintenance, not to mention the problem of finding authors to contribute to the library or who will pay for all this.

In this note we are concerned with the complex problem of maintenance of a repository like MML over time. Such data bases are too complex to even hope that they are designed only once, are built and then grown and are used forever. They evolve and in the case of MML we have

- MML is based on the MIZAR language which albeit slowly but continually evolves. A substantial change to the language may prompt a comprehensive part of MML to be rewritten in order to employ the new language features.
- The MIZAR verifier is continually improved and in order to take advantage of its new possibilities the stored articles have to be updated.
- The organization of the material stored in MML is not fixed. MML can be seen as a collection of intertwined MIZAR articles where authors include whatever is to their liking. Such articles correspond to *primary* scientific information that over time gives rise to the *secondary* information, i.e., overviews, monographs, textbooks. Recently, the MIZAR team has begun building an Encyclopedia of Mathematics in MIZAR, EMM, whose articles have monographical character and are extracted from the "raw" material of the contributed articles in a *semi-automatic* way. This process is assisted by retrieval tools of Grzegorz Bancerek.

At any point, even when we neglect the evolving nature of MML, we may investigate the quality of its contents from the viewpoint of: repetitions of theorems, overlap of defined notions, proper exploration of introduced notions, full usage of the available features of the language, etc. The collection of such qualities we call **integrity**[2] of the stored material. While we would have a hard time giving a succinct definition of the term in this context, we can offer an abundance of excerpts from MML which illustrate violations of integrity and ways to reinstate it.

---

[2] From OED: **integrity 1.** The condition of having no part or element taken away or wanting; undivided or unbroken state; material wholeness, completeness, entirety. **2.** The condition of not being marred or violated; unimpaired or uncorrupted condition; original perfect state; soundness. ...

Integrity of MML is not an absolute notion as it is always relative to the MIZAR language and the MIZAR verifier. Whenever they change, we face the problem of reinstating the integrity and thus we speak of the *evolutionary integrity*. Maintenance of the evolutionary integrity manifests itself in frequent revisions of MML, however, many revisions are caused by other factors as integrity is not the only measure of the quality of MML.

The most interesting are the integrity issues at the semantic level and we will spend most of our time on them; however, we face similar problems at the syntactical and even lexical levels.

2   Lexical Integrity

The set of MIZAR keywords can only be changed when the language changes but otherwise the MIZAR lexicon is not fixed as authors can add new symbols for operators and these symbols are essential for parsing. The symbols are defined in vocabularies and articles import needed vocabularies. One would think that in order to maintain the integrity of symbols it would be enough to check for their repetitions, and this is indeed done. However, there is a bit more to it and when introducing symbols one should foresee possible troubles. Here are two examples.

The frequently used predicate symbol c= from vocabulary HIDDEN for the subset relationship, traps some people. While it is a symbol, it is properly recognized only when the initial c cannot be tokenized as a part of an identifier. So, in ac=a we do not state the case of reflexivity of subset, a c= a, we state the equality of two sets, ac = a. In this case though, the advantages of the succinct notation outweigh potential problems.

Vocabulary CAT_4 introduces a symbol [x] later used in the notation for a functor constructing objects in product categories. Please note that x in this symbol stands for itself and not for any argument. When defining schemes we can use second order predicate variables. To say that a second order predicate variable named P is true for some x we write P[x] (tokenized P [x]). And anyone using this syntax after inserting the CAT_4 into the vocabulary directive gets a bunch of errors as now [x] is tokenized as one symbol. The problem is avoided by using some other identifier than x or using spacing like P[x], or P[x], or P[x]. This is an example where the integrity of the lexicon is violated; introducing symbols like [x] is just a mistake.

3   Syntactic Integrity

When we speak of MIZAR syntax we also consider resolution of symbols and identifiers and not just the context-free part. Notations in MIZAR can be overloaded and this overloading can be easily, albeit involuntarily, abused.

MIZAR offers a number of definitional facilities. In most cases, a definition defines a new constructor of a notion and gives its syntax and meaning. And so, predicates are constructors of atomic formulae, functors are term constructors, modes are type constructors, etc. The syntactic *format* of a constructor specifies the symbol of the constructor and the place and number of arguments. The symbol can be placed in prefix, infix, or postfix position and can take numerous arguments. The format of a constructor together with the information about the types of arguments is called a *pattern* of the constructor. Format is used for context free parsing and pattern for identifying constructors. A pattern together with the identified constructor is called *notation*. A constructor may be represented by different patterns as synonyms and antonyms are allowed. The same pattern can be used for different constructors and this, if abused, renders syntactic analysis almost impossible.

One approach to solve the problem is to require that every pattern denotes a unique constructor or that constructors have unique names. While the former seems overly rigorous, the latter would require a modification of the MIZAR language. The problem can be sometimes resolved using a feature of the MIZAR language. As an example consider the following two definitions

```
let f be Function;
func "f -> Function means                        :: FUNCT_3:def 2
  dom it = bool rng f & for Y st Y in bool rng f holds it.Y = f"Y;

let X, Y be set, f be Function of X, Y;
func "f -> Function of bool Y, bool X means       :: PSCOMP_1:def 2
  for y being Subset of Y holds it.y = f"y;
```

Each of these definitions defines a constructor. The constructors are different and also have different patterns. Note that these definitions coincide for functions which are onto. Consider further a g of type Function of X, Y in an environment where both of the above constructors and notations are available. Because the type of g widens to Function we are in trouble with "g as it is not clear which of the constructors should be identified. The problem can be resolved by writing "(g qua Function) which excludes the second constructor and only the first can be identified. However, which constructor will be identified for "g depends on the order of imports of notations for the above constructors as this order affects the identification process. In this case, if the order of imports is FUNCT_3 and then PSCOMP_1 then "g will be always identified with the second constructor.

4   Integrity of Notions

We present a number of cases where the integrity of MML notions is unsatisfactory and some suggestions of possible techniques to improve the situation.

4.1   Typing Hierarchy

The mode for a *function from a set into a set* is defined in [8, FUNCT_2]

```
let X, Y be set;
mode Function of X,Y is quasi_total Function-like Relation of X,Y;
```

where

```
let X, Y be set;
mode Relation of X,Y  means                              :: RELSET_1:def 1
 it c= [:X,Y:];
```

```
let X, Y be set; let R be Relation of X,Y;
attr R is quasi_total means                              :: FUNCT_2:def 1
 X = dom R if Y = {} implies X = {}
    otherwise R = {};
```

Two other notions are closely related to this notion: a *function from a set* and a *function into a set*. Their MIZAR counterparts are

```
let I be set;
mode ManySortedSet of I -> Function means                :: PBOOLE:def 3
  dom it = I;
```

```
let X be set;
mode ParametrizedSubset of X -> Relation means           :: TREES_2:def 9
  rng it c= X;
```

If f is Function of X, Y for non empty X and Y, then we can demonstrate

```
rng f c= Y; then f is ParametrizedSubset of Y by TREES_2:def 9;
dom f = X by FUNCT_2:def 1; then f is ManySortedSet of X by PBOOLE:def 3;
```

However, while in the current MIZAR these claims have to be proven, they should be available through automatic processing of types. There is no easy remedy and to achieve the desired integrity one would have to change the MIZAR language.

4.2   Different Incarnations of the Same Notion

Ordered pairs. It seems that in a well organized data base there should be one, widely used definition of any commonly needed notion, e.g., a pair of things. At the moment there are three notions of an ordered pair in MML:

1. [x,y] defined as equal { {x, y}, {x} }, that is the Kuratowski pair;
2. <*x, y*> defined as a FinSequence (that is a finite sequence) of length 2 ([6, FINSEQ_1]) extensionally equal to { [1, x], [2, y] };
3. <%x, y%> defined as a finite T-Sequence ([26, AFINSQ_1], see also p. 167) extensionally equal to { [0, x], [1, y] }.

The prevailing opinion is that only one of them should be widely used: preferably the second or the third, while the first should be avoided. The second and third approaches offer natural extensions toward general *n*-tuples but they differ in finite sequences being numbered from 1 or from 0 (see p. 167). Note that at the moment the first definition is used when defining relation, which is used to define function, which is needed to define finite sequence. Therefore the first definition cannot be eliminated so easily.

If we were a bit more ambitious and aimed at an abstract notion of a pair then we would have to define a structure like

```
let S be set;
struct (1-sorted) PairStr over S (# carrier -> set,
   Pair -> Function of [: S, S :], the carrier,
   Left, Right -> Function of the carrier, S        #)
```

and then for a pair we would use `Element of P` where P is a `PairStr` over some set. Note however, that such a definition uses the notions of a Cartesian product and a function, but in order to define them we need some notion of pairing.

So maybe we are doomed to use more than one notion of pairing after all.

Products. When dealing with products of families of sets we face a similar situation. In MML we have

- The binary Cartesian product is defined in [7, ZFMISC_1] as a set of pairs.

```
let X1,X2;
func [: X1,X2 :] means                         :: ZFMISC_1:def 2
   z in it iff ex x,y st x in X1 & y in X2 & z = [x,y];
```

- The product of an arbitrary family of sets is defined in [2, CARD_3].

```
let f be Function;
func product f -> set means                    :: CARD_3:def 5
   x in it iff ex g being Function st x = g & dom g = dom f &
      for x st x in dom f holds g.x in f.x;
```

If we want a Cartesian square, we can use the above as `[:X, X:]` or `product <*X, X*>`; the latter has a semantically equal counterpart available through a different notation: `2-tuples_on X` (see [9, FINSEQ_2]).

It is a bit surprising that despite some efforts at unifying the two notations (see [3, FUNCT_6]), they have not been unified completely. It turns out that trying to work with `product <*X, Y*>`, that is a product of a sequence of sets with length 2, needs quite an investment for developing additional constructors corresponding to the ones already existing for `[: X, Y :]`. Nonetheless, the prevailing opinion is that it is the former that should be widely used.

Similarly to the product of two sets we also have the product of two relational structures, `RelStrs` [13, YELLOW_3] or lattices with topology when we have to deal with several different products at the same time [5, WAYBEL26], and history repeats itself as we have a binary product and a general product. Using both of them causes repetition of essentially equivalent material. This situation indicates the kind of housekeeping work needed for maintaining the integrity of a knowledge base for formal mathematics.

Finite sequences. The type `FinSequence` for finite sequence was introduced into MML in 1989 in [6, FINSEQ_1] as a function whose domain is defined with the help of `Seg n`, the initial segment of natural numbers equal to `{ k : 1 <= k & k <= n }`, [6, FINSEQ_1:def 1] and thus we have finite sequences that are indexed from 1. This was criticized many a time and finally in 2001, finite sequences indexed from 0 were added in [26, AFINSQ_1] as a shorthand for `finite T-Sequence`, where `T-Sequence` [1, ORDINAL1] is a mode of functions having an ordinal as their domain.

Now, a decision needs to be made as to which of the two versions should be kept, as maintaining both seems like a waste of effort. The latter version seems more satisfying while much more machinery has been developed for the former.

4.3   Putting Two Notions Together

The structures used to define metric spaces and topological spaces are introduced in [12, METRIC_1] and [19, PRE_TOPC], respectively

```
struct(1-sorted) MetrStruct (#
                carrier -> set,
                distance -> Function of [:the carrier,the carrier:],REAL #);
```

```
struct(1-sorted) TopStruct (#
                carrier -> set,
                topology -> Subset-Family of the carrier #);
```

We get a metric space by imposing the usual conditions on the distance. The n dimensional Euclidean space is defined in [11, EUCLID]

```
    func Euclid n -> strict MetrSpace equals              :: EUCLID:def 7
      MetrStruct (# REAL n, Pitag_dist n #);
```

with n-tuples_on REAL as the carrier and the Pythagorean distance given by [11, EUCLID:def 5] (see p. 169) as the metric. Now we would like to have a topological space created out of a metric space. This is achieved with the functor

```
    let PM be MetrStruct;
    func TopSpaceMetr PM -> TopStruct equals              :: PCOMPS_1:def 6
      TopStruct (# the carrier of PM, Family_open_set(PM) #);
```

where open sets are defined in terms of Ball. The topological, n dimensional real space is then defined

```
    func TOP-REAL n -> strict TopSpace equals             :: EUCLID:def 8
      TopSpaceMetr (Euclid n);
```

but in doing so we have sort of lost the fact that TOP-REAL n is a Euclidean space. That is, we have not lost it entirely as it can be recovered from the definitions with some workarounds. However, the point is that frequently we want to discuss TOP-REAL n as a topological space and at the same time treat it as Euclid n. It seems that the appropriate solution is to introduce a new structure by multiple derivation from both MetrStruct and TopStruct.

4.4   Repetitions

Repetition of theorems, even with minor variations, while undesirable seems relatively harmless. Repetition of notions causes substantial chunks of the repository to be repeated while elaborating essentially the same material. Repetitions of notions should probably be classified as an error rather than as lack of integrity.

Gluing catenation. For a finite sequence p and two natural numbers m and n, (m, n)-cut p results in a finite sequence being a subsequence of p from index m to n. The so called gluing catenation for finite sequences is defined in [17, GRAPH_2]

```
let p, q be FinSequence;
func p ^' q -> FinSequence equals            :: GRAPH_2:def 2
 p^(2, len q)-cut q;
```

The gluing catenation `^'` catenates two finite sequences removing the first element from the second argument, if possible. Half a year after the above appeared in MML, the following definition was included in [4, REWRITE1]

```
let p,q be FinSequence;
func p$^q -> FinSequence means              :: REWRITE1:def 1
 it = p^q if p = {} or q = {}
 otherwise ex i being Nat, r being FinSequence
         st len p = i+1 & r = p|Seg i & it = r^q;
```

This catenation, denoted $\$^\wedge$, also catenates two finite sequences removing the last element from the first argument if possible.

Both these functors are meant to be used when the last element of the first sequence is the same as the first element of the second sequence; for practical purposes they are equivalent and one of them should be eliminated.

It is interesting to note that only an analogue of the first functor can be defined for transfinite sequences, and, should the finite sequence be introduced as finite transfinite sequence (see p. 167), then the second of the above functors probably would have never seen the light of day.

Norms. The original norm for topological real space is defined in [23, TOPRNS_1]

```
let N be Nat; let x be Point of TOP-REAL N;
func |.x.| -> Real means                    :: TOPRNS_1:def 6
 ex y be FinSequence of REAL st x = y & it = |.y.|;
```

The following definition appeared much later

```
let n be Nat; let p be Point of TOP-REAL n;
func |. p .| -> Real means                  :: JGRAPH_1:def 5
 for w being Element of REAL n st p = w holds it = |.w.|;
```

Here `Element of REAL n` widens to `FinSequence of REAL` and `Point of TOP-REAL n` is also a `FinSequence of REAL` and thus both these definitions are equivalent (the second definition is bound to disappear in the next revision of MML). Yet, in the second of these articles we find the same series of theorems that were proven in the first one. There might be, however, conceivable reasons that someone prefers the definiens of the latter definition, but then it should have been introduced in a redefinition

```
let n be Nat; let p be Point of TOP-REAL n;
redefine func |. p .| -> Real means
 for w being Element of REAL n st p = w holds it = |.w.|;
```

which would render many theorems in JGRAPH_1 redundant. This is because such a redefinition does not introduce a new notion; it only adds another equivalent definiens to the existing constructor. A current revision of MML is introducing this redefinition and eliminating a whole bunch of theorems. Even with this change, complete integrity will not be achieved because in EUCLID we already have

```
let x be FinSequence of REAL;
func |.x.| -> Real equals                              :: EUCLID:def 5
  sqrt Sum sqr abs x;
```

and thus even `TOPRNS_1:def 6` should have been a redefinition of the above. This would be troublesome to get in the current version of MIZAR. While a `Point of TOP-REAL n` is a `FinSequence of REAL`, the type `Point of TOP-REAL n` does not widen automatically to `FinSequence of REAL`. In order to achieve the desired result, one would have to introduce a notion of a finite function into reals and use it as the basic notion for defining norms.

5   Formulating Theorems

In [10, `PSCOMP_1:8`] we read

```
for X being non empty Subset of REAL st X is bounded_below holds
  r = inf X iff (for p st p in X holds p >= r) &
                (for q st (for p st p in X holds p >= q) holds r >= q);
```

We will argue that formulating such equivalences as two separate implications allows us to see things in a better light. Instead of the above we could have

```
Th1:
for X being non empty Subset of REAL st X is bounded_below & r = inf X
 holds (for p st p in X holds p >= r) &
       (for q st (for p st p in X holds p >= q) holds r >= q);
Th2:
for X being non empty Subset of REAL
 st X is bounded_below & (for p st p in X holds p >= r) &
   (for q st (for p st p in X holds p >= q) holds r >= q);
 holds r = inf X
```

The first fact should be rather written as

```
Th1': for X being non empty Subset of REAL st X is bounded_below
       holds (for p st p in X holds p >= inf X) &
             (for q st (for p st p in X holds p >= q) holds inf X >= q);
```

But this is really two theorems bundled together, so they had better be split

```
Th1'1: for X being non empty Subset of REAL st X is bounded_below
        holds for p st p in X holds p >= inf X;
```

```
Th1'2: for X being non empty Subset of REAL st X is bounded_below
        holds for q st (for p st p in X holds p >= q) holds inf X >= q;
```

We learn from [10, `PSCOMP_1`] that `inf X` is just a synonym of `lower_bound X`, the latter defined as

```
let X; assume (ex r st r in X) & X is bounded_below;
func lower_bound X -> real number means               :: SEQ_4:def 5
  (for r st r in X holds it <= r) &
  (for s st 0 < s ex r st r in X & r < it+s);
```

Therefore, `Th1'1` is an immediate consequence of the definition and thus has no room in any data base aspiring toward integrity. On the other hand, the fact labeled `Th1'2` should be kept as an important theorem. Continuing minor improvements, we note that the fact labeled `Th2` is better expressed by

```
for X being non empty Subset of REAL
  st (for p st p in X holds p >= r) &
     (for q st (for p st p in X holds p >= q) holds r >= q);
  holds r = inf X
```

because of the definition of `X is bounded_below` in [14, `SEQ_4:def 2`] which means that `ex p st for r st r in X holds p <= r`. There is no point in saying the same thing twice among the premises.

Achieving integrity is not only the matter of proper initial design; we have to strive for it as the system evolves. Here are two cases in point. In a much later article, [16, `JORDAN5D:1`] we have

```
for B being Subset of REAL, s being Real
  st (ex r be Real st r in B) &
     B is bounded_below & for r being Real st r in B holds s <= r
  holds s <= lower_bound B;
```

The first condition just states that `B` is non empty and it is better expressed by `B <> {}` or with the adjective `non empty` (see Section 6-3). The second condition is spurious in the presence of the third for the reasons given above. But what is then left after these changes is just `Th1'2` that we stated above. The conclusion is simple, if `PSCOMP_1` was written with integrity in mind, then probably `JORDAN5D:1` would have never been introduced.

The most dramatic changes in formulations of theorems are related to the evolution of the MIZAR language and the checking software. The original [10, `PSCOMP_1`, 1997] introduced a number of unary functors for which theorems like the following were proven

```
for X being Subset of REAL holds --X = X;
for X being without_zero Subset of REAL holds Inv Inv X = X;
for X being set, f being Function of X, REAL holds --f = f;
```

All of these theorems state that the respective functor is an involution. In 1997, there was no better way to express such facts. In the meantime, MIZAR has evolved, and now, when introducing a unary functor, we can state that such a functor has the desired property (see [18]), for example

```
let X be Subset of REAL;
func -X -> Subset of REAL equals { -r : r in X};
involutiveness;
```

We have to prove that the functor is indeed an involution but then the verifier tacitly takes this fact into account and there is no need to form the above theorems at all. The introduction of such properties like `involutiveness` prompted a revision of MML in which many theorems were eliminated.

6   Various Rough Edges

1. In definition EUCLID:def 5 (p. 169), it is quite puzzling that abs x is used, as the term is then squared rendering taking the absolute value spurious.
2. In definition FUNCT_2:def 1 (p. 165), the definiens is written in a rather obscure way and the following would be cleaner

```
let X, Y be set; let R be Relation of X,Y;
attr R is quasi_total means                          :: FUNCT_2:def 1
  X = dom R if  Y <> {}
            otherwise R = {};
```

as when X is empty we have that dom R is empty and thus R is empty as well.
3. In several theorems in article [14, SEQ_4, 1989], we see something like

```
for X being Subset of REAL
  st (ex r being real number st r in X) & ... holds ...
```

which has a much cleaner formulation as

```
for X being non empty Subset of REAL st & ...  holds ...
```

However, in 1989 there were no attributes in MIZAR and the cleaner formulation was not an option.
4. Associated with finite sequence is the choice of terminology for the length of such a sequence. The length of a FinSequence is defined as a synonym to its cardinality in [6, FINSEQ_1]

```
let p be FinSequence;
redefine func Card p -> Nat means                    :: FINSEQ_1:def 3
  Seg it = dom p;   synonym len p;
```

Frequently, in theorems about finite sequences we have to say that some index i is in the domain of the sequence p, and this can be done in at least three ways

```
        i in dom p, or i in Seg len p, or  1 <= i & i <= len p
```

Although translations between these representations are trivial, we have the following question: which of them should be used? Individual tastes differ and thus all three are being used. However, this causes major headaches. For instance, theorem [25, FINSEQ_3:27] was formulated, stating that

```
        n in dom p iff 1 <= n & n <= len p;
```

and it is referenced at least 2397 times in MML.

7 Conclusions

Maintaining a repository of formally checked mathematics poses a number of challenges. We attempted to illustrate one of these challenges that we christened *maintaining integrity*. In our opinion such a data base can be only developed in an evolutionary way with numerous contributors. Thus maintaining the integrity of the data base becomes a serious issue. While we can give numerous examples of integrity violations we are still looking for general mechanism that would assist us in achieving the integrity of the MIZAR Mathematical Library.

Numerous examples presented in this note indicate that the issues of integrity must be considered with respect to the checking software and features of the language. Integrity is not an absolute notion.

References

1. Grzegorz Bancerek. The ordinal numbers. *Formalized Mathematics*, 1(**1**):91–96, 1990.
2. Grzegorz Bancerek. König's theorem. *Formalized Mathematics*, 1(**3**):589–593, 1990.
3. Grzegorz Bancerek. Cartesian product of functions. *Formalized Mathematics*, 2(**4**):547–552, 1991.
4. Grzegorz Bancerek. Reduction relations. *Formalized Mathematics*, 5(**4**):469–478, 1996.
5. Grzegorz Bancerek. Continuous lattices of maps between T_0 spaces. *Formalized Mathematics*, 9(**1**):111–117, 2001.
6. Grzegorz Bancerek and Krzysztof Hryniewiecki. Segments of natural numbers and finite sequences. *Formalized Mathematics*, 1(**1**):107–114, 1990.
7. Czesław Byliński. Some basic properties of sets. *Formalized Mathematics*, 1(**1**):47–53, 1990.
8. Czesław Byliński. Functions from a set to a set. *Formalized Mathematics*, 1(**1**):153–164, 1990.
9. Czesław Byliński. Finite sequences and tuples of elements of a non-empty sets. *Formalized Mathematics*, 1(**3**):529–536, 1990.
10. Czesław Byliński and Piotr Rudnicki. Bounding boxes for compact sets in $]^2$. *Formalized Mathematics*, 6(**3**):427–440, 1997.
11. Agata Darmochwał. The Euclidean space. *Formalized Mathematics*, 2(**4**):599–603, 1991.
12. Stanisława Kanas, Adam Lecko, and Mariusz Startek. Metric spaces. *Formalized Mathematics*, 1(**3**):607–610, 1990.
13. Artur Korniłowicz. Cartesian products of relations and relational structures. *Formalized Mathematics*, 6(**1**):145–152, 1997.
14. Jarosław Kotowicz. Convergent real sequences. Upper and lower bound of sets of real numbers. *Formalized Mathematics*, 1(**3**):477–481, 1990.
15. MIZAR *Manuals*. http://mizar.org/project/bibliography.html.
16. Yatsuka Nakamura and Adam Grabowski. Bounding boxes for special sequences in $]^2$. *Formalized Mathematics*, 7(**1**):115–121, 1998.
17. Yatsuka Nakamura and Piotr Rudnicki. Vertex sequences induced by chains. *Formalized Mathematics*, 5(**3**):297–304, 1996.

18. Adam Naumowicz and Czesław Byliński. Basic Elements of Computer Algebra in MIZAR. *Mechanized Mathematics and Its Applications*, **2**(1):9–16, 2002.

19. Beata Padlewska and Agata Darmochwał. Topological spaces and continuous functions. *Formalized Mathematics*, 1(**1**):223–230, 1990.

20. P. Rudnicki, Ch. Schwarzweller and A. Trybulec. Commutative Algebra in the Mizar System. *Journal of Symbolic Computation*, **32**:143–169, 2001.

21. Piotr Rudnicki and Andrzej Trybulec. On equivalents of well-foundedness. *Journal of Automated Reasoning*, **23**(3-4):197–234, 1999.

22. Piotr Rudnicki and Andrzej Trybulec. Mathematical Knowledge Management in MIZAR. *1st Int. Workshop on MKM*, Sept. 24-26, 2001, http://www.risc.uni-linz.ac.at/institute/conferences/MKM2001.

23. Agnieszka Sakowicz, Jarosław Gryko, and Adam Grabowski. Sequences in $]_t^N$. *Formalized Mathematics*, 5(**1**):93–96, 1996.

24. Andrzej Trybulec. Tarski Grothendieck set theory. *Formalized Mathematics*, 1(**1**):9–11, 1990.

25. Wojciech A. Trybulec. Non-contiguous substrings and one-to-one finite sequences. *Formalized Mathematics*, 1(**3**):569–573, 1990.

26. Tetsuya Tsunetou, Grzegorz Bancerek, and Yatsuka Nakamura. Zero-Based Finite Sequences. *Formalized Mathematics*, 9(**4**):825–829, 2001.

27. Freek Wiedijk. MIZAR: *An Impression.* http://www.cs.kun.nl/~freek/notes.

A Theoretical Analysis of Hierarchical Proofs

Paul Cairns and Jeremy Gow

UCL Interaction Centre, University College London,
26 Bedford Way, London WC1H 0AP, UK
{p.cairns,j.gow}@ucl.ac.uk
www.uclic.ucl.ac.uk/imp

Abstract. Hierarchical proof presentations are ubiquitous within logic and computer science, but have made little impact on mathematics in general. The reasons for this are not currently known, and need to be understood if mathematical knowledge management systems are to gain acceptance in the mathematical community. We report on some initial experiments with three users of a set of web-based hierarchical proofs, which suggest that usability problems could be a factor. In order to better understand these problems we present a theoretical analysis of hierarchical proofs using *Cognitive Dimensions* [6]. The analysis allows us to formulate some concrete hypotheses about the usability of hierarchical proof presentations.

1 Introduction

Rigorous proof is a central aspect of modern mathematics. In providing a coherent organisation of mathematical knowledge, any human-oriented system must consider not only how to organise, reference and retrieve proofs but also how to present the retrieved proofs to the working mathematician. Hierarchical proof presentations have a long history in logic and computer science but have received little attention in the wider mathematical community, despite having a number of advocates [12,13].

Hierarchical proofs seem to be an ideal technique for the presentation of digital mathematics. Firstly, their modular structure makes them suitable for interactive, hypertext presentations. Secondly, there is already some experience of presenting hierarchical proofs within the theorem proving systems community. Last but not least, they make explicit the 'true structure' of a proof which is suppressed in textbook, linear presentations, e.g. [9]. However, a key question is still unanswered: whether they help the reader to understand the proof, compared to a traditional presentation.

Doubts about the usability of hierarchical proofs were raised by a small pilot study of the Polya-Lamport framework [2], a presentation framework for mathematical texts, which we present in Section 3. Unfortunately, this leaves us with few definite hypotheses which might form the basis of further experiments. We look to theory to help — specifically, Cognitive Dimensions [6,5], a lightweight evaluation technique for the usability of interactive devices and notations. In Section 4, we use it to provide a theoretical analysis of hierarchical proofs. Based

A. Asperti, B. Buchberger, J.H. Davenport (Eds.): MKM 2003, LNCS 2594, pp. 175–187, 2003.

on this and the pilot study, Section 5 contributes some testable ideas on the usefulness and usability of hierarchical presentations of mathematical proofs.

It is important to stress that the contribution of this paper is not to provide evidence for or against the usability of hierarchical proofs, but to generate some concrete hypotheses that will form the basis of more rigorous experimental tests.

2   Hierarchical Proofs

It has long been recognised within the mathematical community that rigorous logical proofs can be represented hierarchically. That is, since statements of a proof depend logically on axioms or previous statements in the proof, the structure of the proof can be understood as a directed acyclic graph (or tree). However, in practice, the full logical structure is rarely presented, or even fully considered, and most published proofs take a sequential form like most printed material, e.g. [17,8]. A few explicit logical links are made through the numbering of lemmas and definitions and the labeling of formulae rather than by any graphical means. Most links are left implicit, to be reconstructed by the reader.

Research into proofs with an explicit hierarchical structure falls into two broad categories: attempts to introduce hierarchical proofs into mathematics, and work on logic, including theorem provers and related systems.

Lamport [12] proposed a standard for presenting hierarchical proof structure that employs indentation and labeling of proof statements, without resorting to graphical methods. A proof is made up of an optional list of labeled statements, each with its own proof, followed by a list of references to statements and external theorems and definitions. Thus, a proof may be broken down progressively into finer and finer detail until the bottommost steps are trivial. Further annotations are used to indicate definitions and case analyses within the proof. An example proof is given in Figure 1. Lamport recommends the hierarchical format, as it 'makes it much harder to prove things that are not true'. He also notes its suitability for hypertext systems.

Lamport's approach has been implemented in HTML for calculational style proofs (i.e. linear derivations) [7]. The hierarchical nature of the proof gives a natural way to hide and reveal details in a proof: substeps in the proof can be hidden or revealed on a mouse click.

An alternative way of hierarchically structuring proofs was proposed by Leron [13]. Whereas Lamport emphasises the deductive structure of a proof, Leron advocates using a hierarchy to highlight the conceptual structure. The top level of the proof explains the key ideas and proof outline, with lower levels making the ideas more concrete and filling in the details, until the author has nothing more to add. Furthermore, Leron's format is less structured, with each node[1] of the hierarchy corresponding to a collection of paragraphs, rather than an individual assertion. The proof presentation also incorporates heuristic and informal knowledge, with 'elevator' sections discussing the transition between levels.

---

[1] Confusingly, he calls nodes 'levels'. Hence they may be several 'levels' of depth 2, called 2.1, 2.2,...

PROOF SKETCH: Given a finite set of triadic primes, we construct a number that has a triadic prime factor not in the set.

GIVEN: There are only finitely many triadic primes $\{p_1, \ldots, p_n\}$

THEN: A contradiction

⟨1⟩1. 3 is a triadic prime, say $p_1 = 3$
   PROOF: By definitions of triadic and prime.
LET: $M \doteq 4p_2 \ldots p_n + 3$
⟨1⟩2. No $p_1, p_2, \ldots, p_n$ divides M
   ⟨2⟩1. p_1 does not divide $4p_2 \ldots p_n$
   ⟨2⟩2. $p_2, \ldots, p_n$ have remainder 3 from M
   ⟨2⟩3. Q.E.D.
     PROOF: By ⟨2⟩1 and ⟨2⟩2.
⟨1⟩3. M has a triadic prime factor
GIVEN: M's prime factors are monadic
THEN: A contradiction
   ⟨2⟩1. A product of monadic numbers is monadic
   ⟨2⟩2. M is monadic
   ⟨2⟩3. Q.E.D.
     PROOF: By ⟨2⟩2 and definition of M.
⟨1⟩4. Q.E.D.

Fig. 1. A partial Lamport-style proof of the infinity of triadic primes.

The difference in emphasis can be seen by comparing Figures 1 and 2. Both are proofs of the infinity of 'triadic primes', defined by Leron as these with remainder 3 on division by 4. Lamport's proof hierarchy draws attention to the argument structure, and omits heuristic information. Leron's proof hierarchy highlights the key ideas and refines them. Individual levels are written in a traditional proof style.

Amongst the automated and interactive theorem proving community, there are numerous examples of user interfaces that use a hierarchical representation. XIsabelle was developed as an interface for the Isabelle interactive proof system [3]. A goal is graphically linked to its subgoals, forming a tree structure. Interactive proof is supported by letting the user select goals from the tree and apply further actions to them.

Similar presentation principles can, and have been, applied to other theorem provers and proof-oriented systems. For instance, the ΩMEGA and λClam proof planners both have graphical user interfaces that can represent proofs as trees, respectively, LΩUI [18] and XBarnacle [11]. Proofs in proof planning are not always proofs in a particular logic, but rather a tree of *proof method* applications. Most proof plans are considerably more logically rigorous than most mathematicians, so deserve the term proof in this context.

Another example is the JAPE system, which is intended as a tool for teaching logic to computing science students [1]. As such it is able to work in a variety of logics and proof presentations. One of the presentations is like the paper notation of tableau proofs. Another is more text-based, but the proof of a step

Level 1 Suppose the theorem is false and let $p_1, p_2, \ldots, p_n$ be all the triadic primes. We construct in (Level 2) a number M have the following two properties:

(a) M as well as its factors are different from $p_1, p_2, \ldots, p_n$;
(b) M has a triadic prime factor.

These two properties clearly produce a contradiction, as we get a triadic prime which is not one of $p_1, p_2, \ldots, p_n$. Thus the theorem is proved.

In The Elevator How shall we approach the definition of M? In light of Euclid's classical proof, it is natural to try $M = p_1 p_2 \ldots p_n + 1$. This indeed...

Level 2 Let $M = 4p_2 \ldots p_n + 3$ (we assume $p_1 = 3$). We show that M satisfies the two requirements from Level 1. Requirement (a) means...

Level 3 Lemma: A product of monadic numbers is again a monadic number.

Fig. 2. Extracts from a Leron-style proof of the infinity of triadic primes.

has a subproof made up of substeps and so on. In both cases though, the interface presents the logical connection between steps and allows users to select steps for further development of the proof.

Thus, there are many ways in which hierarchical proofs have been used both in mathematics and in computer science. Yet despite the naturalness and obviousness of the representation, it is notable that none have wide use in the general mathematical community. This suggests that there is some factor working against the use of hierarchical presentations.

3   Initial Experiments

We developed the Polya-Lamport framework as part of an investigation into using interaction to improve presentations of mathematical texts. The 'Polya part' of the framework built on the heuristic problem solving approach proposed by Polya [16] to provide users with structured relationships to the wider context of a particular result or definition. The 'Lamport part' was based on the method of structuring proofs to provide users with control over the amount of the detail they see in a proof. We will only discuss the actual proof presentation here. The full framework is described elsewhere [2].

3.1   The Web-Based Hierarchical Proofs

The presentation of proofs in our framework is very similar to that proposed by Lamport. A proof is made up of a sequence of steps and each step may have its own proof made up of further steps. The proof is preceded by a list of 'background' results and definitions. Though Lamport only demonstrated the structure of proofs through examples, it was relatively easy to formalise the structure into a grammar for an XML DTD. Thus, all proofs were actually written by hand in XML and converted using XSLT to a combination of HTML, GIFs and JavaScript that could be viewed on the two most common browser platforms, Netscape and Internet Explorer. (See [20] for information on these various web standards.)

The pages provided two ways to interact with a proof. First, statements that had their own proofs had squares next to them containing + if the subproof was hidden or − if it was revealed. Clicking the squares toggled between the states. Hidden subproofs were summarised as a list of proof steps and external results used in the proof, e.g. "By assumption 2 and background 3."

Secondly, each reference to a proof step or background result/definition was a hyperlink. Clicking the link moved the proof display to the corresponding part of the proof. Hovering over the link displayed the target statement in a separate frame at the bottom of the page[2], without altering the main proof display. Thus, the proof presentation was a combination of common web and user interface interactions built into Lamport's proof structure.

The new presentation was intended to be compared with more traditional proof presentations on paper. Thus, the content of the hierarchical proofs was made to resemble the normal language of the mathematical vernacular [10]. Logically necessary information was often omitted because it was irrelevant or could be easily reconstructed from the context. For example, we say X is a topological space even though this is formally defined as a pair $(X, \mathcal{T})$.

3.2   A Limited User Study

A pilot user trial was used to elicit feedback on the Polya-Lamport framework, including hierarchical proofs. The basic content of the trial was an undergraduate course in topology written by Peter Collins, taken by all first year mathematicians at Oxford University. The course notes were adapted into the Polya-Lamport framework using the original examples, diagrams and language as far as possible. Details of the conversion of the materials is given elsewhere [2]. These formed a set of online materials that were used in the trial.

The online notes were made available to three first year mathematicians at St Edmund Hall, Oxford University, who were currently studying the topology course and hence would have a real motivation to use them. They were interviewed after three weeks. The online materials were available on a lap-top as a prompt during the interview, because users are not always explicitly able to remember software [4]. The interviews were intended to elicit qualitative data on the experience of using the Polya-Lamport framework, including the hierarchical proofs, and were recorded on MiniDisc for subsequent analysis. The results presented here are just the analysis of the online materials that involved the interactive, hierarchical proofs.

Though three users is not enough for any rigorous trial, it is enough to begin a qualitative analysis of the data and to start the formation of a grounded theory as part of a larger investigation [19]. Specifically, we can formulate hypotheses to test in more focused experiments. This is discussed further in Sections 4 and 5.

---

[2] Except in Internet Explorer — as usual, Microsoft's interpretation of a standard (the HTML DOM) differed from everyone else's.

One of the most noticeable outcomes of the interviews were that the users quickly recognised the hierarchical structure even though they had not been primed to think of it as such:

INTERVIEWER: In terms of the proofs you've already mentioned ... what do you think about how they are laid out?

SUBJECT 1: They seem laid out quite well. Very easy to see where it is going.

And even one student who declared himself as "not very good with computers" recognised what was going on:

INTERVIEWER: So you are not used to seeing that [format]?

SUBJECT 2: No, but once I worked it out, it seemed okay.

Labels were used to refer to different parts of the proof within a given proof step and these corresponded to hyperlinks as described earlier. One user thought that this could be confusing. Another user experienced that confusion:

SUBJECT 3: If you need to click that to [go] back to the actual material. I just can't concentrate... to the program.

This wasn't just a feature of the hierarchical proofs. Any links in the framework which took the user away from the focus of their attention were sometimes seen as negative.

As for the hiding and revealing of proof substeps, the response was rather mixed, mainly because of the summaries of the proofs. The summary proofs collected together the statements that the hidden proof steps used, referring to them by their labels. However, this was often viewed as a meaningless list of numbers and, without the actual statements there, tended to be off-putting:

INTERVIEWER: What was difficult about seeing it online? ...

SUBJECT 2: I would like to have... more words.

In addition, because of problems with the XSLT used to generate the summaries, the statements labels came up in the order in which they were used in the proof, sometimes multiple times. This meant that a summary might appear as "By background 3, background 2, background 1" which had a tendency to distract the users:

INTERVIEWER: How did you find the numbering of the theorem? ...

SUBJECT 1: I found it a bit bizarre...

There also seemed to be a deeper problem with the content of the hierarchical proofs themselves. The aim of the hierarchical structure was to reveal the more detailed steps as the reader descended the hierarchy. However, though the principle was recognised it was not understood that way:

SUBJECT 3: It is not very detailed, I think. ... [I found] more details but not very clear.

And:

SUBJECT 2: ...it is not all out there.

It seems that the details were not in a recognisable form or that what was there was not the "whole proof" despite being logically more complete. This gave the tendency to "skim-read" and "go through it fast".

On the positive side though, the hope of giving the user control over how much detail they read was realised to some extent. One user liked it:

SUBJECT 2: It was helpful to shrink it down ... Because you could expand the parts to see exactly what you wanted. It is good to have that flexibility.

In summary then, the user trial gave some initial pointers to problems of hierarchical proofs. Though the structure of the proof is easily recognised and manipulated, it does not seem to be easily understood. Some of this may be due to artificial elements introduced such as labeling, especially in the summaries of proofs. A deeper factor may be that the steps of the proof given do not seem to have the 'right sort' of details even though they are logically more detailed than traditional proofs. It seems to be the quality of the details and not the quantity of the details that is important for the readers. However, given the small and imprecise nature of this trial, this conclusion can only be tentative. We turn to a more theoretical approach to look for further insights.

4   An Analysis over Cognitive Dimensions

Cognitive dimensions were developed by Thomas Green and others to reflect aspects of the usability of information representations that can be independently considered, though in a given situation may be mutually constraining. Where they differ from guidelines and heuristic evaluation [14] is that they consider inherent features of working with an information representation, rather than the more superficial aspects of user interfaces. Moreover, they are motivated and inspired by findings in cognitive science, although the connection is somewhat tenuous.

Given the abstract nature of the dimensions, a considerable amount of interpretation may be required to analyse a particular domain. However, they can still serve as a very useful tool for thinking about the inherent usability issues of some domains, over and above more superficial usability guidelines [14]. A detailed description of the technique and its application to the evaluation of visual programming languages can be found in [6].

In the analysis, we consider both the structure and the interaction of the hierarchical proofs, often by comparison to traditional proof presentations. To avoid the danger of the interview biasing the analysis, the second author performed the analysis before learning either of the findings from the interviews or the first author's analysis. The analyses corresponded closely.

4.1   Applying the Dimensions

In the sections below we analyse hierarchical proofs over a range of cognitive dimensions. Some dimensions are omitted, as they would refer to writing/editing proofs, whereas this paper is concerned with presentation and understanding. Another, role-expressiveness, is also omitted as it is not clear what it means — a view that Green himself also holds [5].

Abstraction Gradient: Abstraction gradient refers to the number of abstract concepts that need to be learnt by the user before they can understand the hierarchical representation of the proof. This is independent of the amount of abstraction used in the proof or the underlying mathematics.

Any hierarchical proof requires some abstraction as the breakdown of the proof into a hierarchy is an abstraction of logical argument into e.g. levels, steps, assumptions. Furthermore, the steps also have individual labels such as GIVEN for assumptions and LET for definitions. These must be learned before the proof can be understood.

However, the abstractions used rely on other notations within either computing or mathematics. The hierarchical structure is relatively common (e.g. Windows Explorer) and the terms used to denote elements of the hierarchy have been taken from mathematics. Thus, though there are some abstractions to be learnt and these can be off-putting to new users, they are not so far removed from common mathematical experience and so should not pose an insurmountable barrier.

Closeness of Mapping: The closeness of mapping is about mapping the representation of the proof to the process of proving. The closer the mapping the easier it is for users to relate the proof to their understanding of the problem. Neither hierarchical nor traditional proof presentations could have a claim to being 'closer to proof' without some cognitive evidence about how users understand the proof process. As with the psychology of programming, this is not a well understood area [6], and we are currently unlikely to find convincing arguments either way.

One might argue that a user's understanding will be shaped by the traditional proof presentations they are used to, but this dodges the questions of whether either presentation style is in some sense better than the other, and whether the user can adapt their understanding.

Consistency: Consistency is the extent to which users can transfer their experience in one part of the proof to the understanding of another. Inasmuch as a hierarchy imposes a regular structure on a proof, hierarchical proofs are highly consistent. Certainly much more so than traditional proofs, which have no generally accepted structure. The consistency of hierarchical proofs could help users to learn the representation and to extract information more quickly from unfamiliar proofs.

Diffuseness/Terseness: Diffuseness and terseness are the quantity of symbols and space needed to represent an element of the hierarchical proof. They are more diffuse than traditional presentations in that i) they use space to indicate the hierarchical structure, and ii) making proof steps and logical dependency explicit increases the amount of text — similar to the 'de Bruijn factor' of formal mathematics [21]. Hiding subproofs can reduce diffuseness, at least until the subproof is revealed. Unfortunately, the diffuseness caused by giving the user more structure and explicit information, may have the effect of separating related information, increasing their cognitive workload (see the visibility dimension below).

In another sense, hierarchical proofs are more terse than traditional presentations, in that when explicit references are given, they use abstract labels, e.g.

"By step (3)2" rather than a description of content, e.g. "p_1 does not divide $4p_2, \ldots, p_n$". This requires the user to hold more of the proof in mind, or to look up the references. Again, this could cause problems.

We could say that hierarchical proofs are 'structurally diffuse' and 'referentially terse', whereas traditional proofs tend towards 'structural terseness' and (when explicit references are made) 'referential diffuseness'.

Hard Mental Operations: Hard mental operations refer to the user having to spend time working to understand the proof presentation. Obviously, a complex mathematical proof is complex no matter how it is presented. Rather, this dimension is concerned with any mental workload placed on the user by the presentation itself. There is some extra mental workload due to having to remember or look up statements given their labels. This is caused by the diffuseness/terseness issues discussed above.

Hidden Dependencies: These are relationships between parts of the proof that are not made visible by the presentation. The relationship of 'substep' is made clear by the hierarchical structure. Also the number of hidden logical dependencies is reduced by requiring them to be explicitly stated. They avoid the traditional implicit argument where very few words can be used to communicate extremely complex steps [15]. Not all relationships are made clear in hierarchical proofs, e.g. the user cannot easily discover where a particular statement is used later in the proof — although the same is true of traditional presentations. This could be amended by enhancing the hierarchical presentation, but there is the risk of unnecessarily complicating it.

Progressive Evaluation: Progressive evaluation is the ability of the user to assess their progress through the proof either in terms of how much they still have to read or how much they still have to understand. For novice users, progressive evaluation is essential and it can still be useful to expert users.

By making the structure explicit, and relegating less important detail to lower levels, the user can evaluate their progress through the proof much more easily. For example, they can see they are reading the second of three sections to the proof. But hiding subproofs may obscure the amount still to be understood within a subproof. This could be relieved by adding a 'reveal all' option, or an indication of the size of the hidden subproof.

Secondary Notation: This addresses the amount to which the presentation can convey additional meaning beyond the actual proof, e.g. discussions about the proof or related diagrams. The hierarchical proofs use secondary notation, namely indentation, to denote the proof structure. Keywords also indicate various types of statement. Thus, these cues should be useful to the reader. Traditional presentations may of course include arbitrary additional information to help the user, but the hierarchical proof format allowed 'proof sketches' to be included that could also play this role. Diagrams are a special case — they were necessarily separate from the structured proof, and thus cannot be used by the reader without a shift of attention.

Visibility and Juxtaposition: Visibility is about how much of a proof can be seen at any one time, that is, without cognitive work. Juxtaposition is allowing different parts of a proof to be placed side by side for comparison and hence makes up a significant component of visibility.

It is possible to reveal all parts of a proof but usually, because of the structure imposed on the proof, it is unlikely that it would all fit in one window. It may be that such a full view of the proof would not actually be useful. A more serious visibility issue is the use of abstract labels, mentioned above. With reference and referent often not both visible, the user is required to remember the referent or to move their attention away from their current focus in order to retrieve it.

Juxtaposition is not built in to the implementation but it would not be a problem for users to re-open the proof in another window and so bring different parts of the proof together. Juxtaposing two parts of the proof is easier if they are within the same proof level — the user can simply close the intervening sub-proofs. Comparison between levels may be harder, especially with the increased diffuseness of the proof relative to traditional presentations.

4.2   Outcome of the Analysis

The good usability features identified by the cognitive dimension analysis are:

- A hierarchy provides a more *consistent* presentation.
- The proof structure helps a user to *progressively evaluate* their journey through the proof.
- Logical *hidden dependencies* are made explicit.
- The *secondary notation* gives clues to structural features of proof, and the roles of some statements.

Features of hierarchical proofs that might cause problems are:

- Introduces several *abstractions* which may take time to learn.
- Increased *diffuseness*, caused by the de Bruijn factor and proof layout. This separates related information.
- Referring to statements by their abstract labels is overly *terse*.
- A diffuse proof with terse references increases the amount of *hard mental operations*.
- Hidden subproofs can prevent *progressive evaluation* of the amount achieved by the user.
- Reference and referent are often not both *visible*.

We don't know enough to evaluate *closeness of mapping* — this depends on how people understand proof.

In summary, the cognitive dimensions analysis has provided several concrete hypotheses about the ways in which hierarchical proofs may be more or less usable than traditional presentations. Experimental research will be required to validate these claims.

5   Conclusions and Future Work

Comparing the analytical findings with the interview findings, it is noticeable that the hierarchical structure was easily recognised by users and that they can select what they want to see. However, the problems with referencing statements within the proof is confirmed by the cognitive dimensions. We suggest that abstract labels are not an effective way to present cross-references over the increased distances of a hierarchical presentation. Traditional presentations try to place related information nearby or use a memorable reference.

The users also seemed to dislike the particular form of hierarchical proofs. The cognitive dimensions analysis suggests an explanation: understanding the structure of a proof requires familiarity with new and abstract concepts — this may put the user off, especially as they do not map closely to the form of traditional proofs. This may go away as they become accustomed to it. However, it may be a deeper problem related to the understanding of logical argument. It may well be the case that readers of traditional proof presentations have the same problem. The hierarchical presentation could be highlighting, rather than causing, the problem.

The analysis suggests problems not specifically identified by the trial: user's need to be able to evaluate their progress through the proof, and hierarchical proofs can both help and hinder this. We could also make explicit more of the hidden dependencies between parts of the proof. Diagrams could be more tightly integrated. Only part of the proof is visible at once, and there is no specific support for juxtaposing two parts. Any or all of these may form part of the usability problems. An important direction for further work is to carry out experiments to validate or invalidate our hypotheses.

We speculate that a more fundamental problem with hierarchical proofs may be that the content of the steps is not appropriate. Our proofs were based on traditional proofs, using similar language and symbols, but there could be an inherent problem with hierarchical presentation. Users may not require structuring of the logical detail in the proofs, but help with what is being *communicated* in the proof.

Of course, the hierarchical structure that we used is just one of several possible, even within the Lamport style. Other proof styles might be more accessible. In particular, our theoretical analysis identifies places where the style can be developed. It would also be interesting to investigate a Leron style approach.

It may be that the usability problems are related to individual learning styles. However, the analysis on cognitive dimensions seems to point away from this. They are cognitively motivated, general guides to usability and they indicate problems similar to those actually found with the users.

The way forward is to make further investigations into how users perceive and work with hierarchical proofs and how they compare with traditional proofs. We are planning to do a larger investigation on this aspect of the Polya-Lamport framework involving both qualitative and quantitative feedback on the usability issues. The analysis in this paper has identified areas of concern and given us specific theories to work with.

Regardless of what might be root cause of the usability problems, the key lesson is that a hierarchical proof is not *a fortiori* the best presentation for proofs. There may be many and complex reasons why they are not desirable at the moment. If it is simply a matter of familiarity then steps must be taken to introduce them earlier in a mathematical career rather than later. In any case, systems for mathematical knowledge management should be cautious about placing them as the main interface between mathematicians and their proofs.

Acknowledgments. Many thanks to Peter Collins and the undergraduate mathematicians at St. Edmund Hall, Oxford University, who took part in the user trials. This work was supported by EPSRC, UK, under grant GR/N29280.

References

1. R. Bornat, B. Sufrin, 'Animating Formal Proofs at the Surface: The Jape Proof Calculator,' *The Computer Journal*, 42(3):177–192, 1999
2. P. Cairns, J. Gow, 'On Dynamically Presenting a Topology Course,' *First International Workshop on Mathematical Knowledge Management*, 2001. Revised version submitted to *Annals of A.I. & Math.*, 2002
3. K. Easthaughffe, 'Support for Interactive Theorem Proving: Some Design Principles and Their Application,' in R. Backhouse (ed.), *Workshop on User Interfaces for Theorem Provers*, p96–103, 1998
4. X. Faulkner, *Usability Engineering*, Palgrave, 2002
5. T. R. G. Green, A. Blackwell, *A tutorial on cognitive dimensions*, www.cl.cam.ac.uk/users/afb21/publications/CDtutSep98.pdf
6. T. R. G. Green, M. Petre, ' Usability analysis of visual programming environments: a "cognitive dimensions" framework,' *J. Visual Languages and Computing*, 7:131–174, 1996
7. J. Grundy, 'A Browsable Format for Proof Presentation,' *Math. Universalis*, 2, 1996, www.ant.pl/MathUniversalis/2/grundy/mu.html
8. P. R. Halmos, *Naive Set Theory*, Springer, 1960
9. A. G. Hamilton, *Logic for Mathematicians, (Revised Edn)*, Cambridge University Press, 1988
10. J. Harrison, 'Formalized Mathematics,' *Math. Universalis*, 2, 1996, www.pip.com.pl/MathUniversalis/2/harrison/jrh0100.html
11. M. Jackson, H. Lowe, 'XBarnacle: Making Theorem Provers More Accessible,' *CADE-17*, p502–506, LNAI 1831, Springer, 2000
12. L. Lamport, 'How to write a proof', *American Mathematical Monthly*, 102(7):600–608, 1994
13. U. Leron, 'Structuring Mathematical Proofs,' *American Mathematical Monthly* 90(3):174–185, 1983
14. J. Nielsen, *Usability Engineering*, Morgan Kaufmann, 1993
15. L. C. Paulson & K. Grabczewski, 'Mechanizing Set Theory,' *Journal of Automated Reasoning*, 17:291–323, 1996
16. G. Polya, *How to Solve It*, Penguin Books, 1990
17. W. Rudin, *Functional Analysis (2nd Edn)*, McGraw-Hill, 1991

18. J. Siekmann, S. Hess, C. Benzmüller, L. Cheikhrouhou, H. Horacek, M. Kohlhase, K. Konrad, A. Meier, E. Melis, M. Pollet, V. Sorge, 'LΩUI: Lovely Omega User Interface,' *Formal Aspects of Computing* 11(3):1–17, 1999
19. A. Strauss, J. Corbin, *Basics of Qualitative Research: Techniques and Procedures for Developing a Grounded Theory*, Sage Publications, 1998
20. W3C Recommendations, `www.w3.org/TR/\#Recommendations`
21. F. Wiedijk, *The De Bruijn Factor*, `www.cs.kun.nl/~freek/factor/`

Comparing Mathematical Provers

Freek Wiedijk

University of Nijmegen

Abstract. We compare fifteen systems for the formalizations of mathematics with the computer. We present several tables that list various properties of these programs. The three main dimensions on which we compare these systems are: the size of their library, the strength of their logic and their level of automation.

1 Introduction

We realize that many judgments in this paper are rather subjective. We apologize in advance to anyone whose system is misrepresented here. We would like to be notified by e-mail of any errors in this paper at <freek@cs.kun.nl>.

1.1 Problem

The QED manifesto [6] describes a future in which all of mathematics is encoded in the computer in such a way that the correctness can be mechanically verified. During the years the same dream has been at the core of various proof checking projects. Examples are the Automath project [17], the Mizar project [15, 23], the NuPRL project [8], and the Theorema project [7]. Recently, the checking of mathematical proofs has become popular in the context of verification of hardware and software: important systems in this area are ACL2 [13] and PVS [18]. Because of this, the field of proof verification currently focuses on computer science applications. The study of formal proof in mathematics is still not widespread.

We have compiled a list of 'state of the art' systems for the formalization of mathematics, systems that one might seriously consider when thinking of implementing the QED dream. We were not so much interested in experimental systems (systems that try out some new idea), as well as in 'industrial strength' systems (systems that are in at least some aspects better at the formalization of mathematics than all other existing systems). We ended up with a list of fifteen systems. For each of these systems we asked a user of the system to formalize the same small theorem: the Pythagorean proof of the irrationality

A. Asperti, B. Buchberger, J.H. Davenport (Eds.): MKM 2003, LNCS 2594, pp. 188–202, 2003.

of $\sqrt{2}$.[1] These fifteen formalizations will be published elsewhere.[2] We did not in advance specify to the user a very specific proof problem to be solved, because we wanted formalizations that were the most natural for the given system.

In this paper we compare these fifteen systems according to various criteria. We do not try to establish which of these systems is 'best': they are too different to say something like that. All of these fifteen systems clearly show the dedication of their creators, they all contain important ideas, and they all merit to be studied and used. The main purpose of this paper is to show how different this kind of system can be. When one only knows a few of these systems, it is tempting to think that all systems for formal mathematics have to be of a similar nature. Instead, it is surprising how diverse these systems are.

1.2 Approach

This paper is primarily a collection of tables that show various properties of the systems. It is something like a 'consumer test'.

To illustrate three of the most important dimensions for comparing these systems (the size of their library, the strength of their logic, and their level of automation), at the end of the paper we show them together in a two-dimensional diagram. We realize that this diagram is highly subjective. We list some aspects of the systems that we have used to determine the positions in this diagram. (Readers will probably disagree with the details of this diagram and are encouraged to make their own variant.)

The order in which we list the systems in this paper is the order in which we received the formalizations of the irrationality of $\sqrt{2}$. In this way we wish to express our gratitude for the help of all the people who wrote these formalizations.

1.3 Related Work

On page 1227 of [4] there appears a comparison similar to the one in this paper. However, it only compares nine instead of fifteen systems, and the comparison takes only one and a half pages.

---

[1] This proof is mentioned in Aristotle's *Prior Analytics*, as follows [10]:

> For all who argue per impossibile *infer by syllogism a false conclusion, and prove the original conclusion hypothetically when something impossible follows from a contradictory assumption, as, for example, that the diagonal [of a square] is incommensurable [with the side] because odd numbers are equal to even if it is assumed to be commensurate.*

It was interpolated in Euclid's *Elements* as Proposition x. 117. Due to the oral tradition of the Pythagorean School, the origins of this proof have been covered in 'complete darkness' [22]. There is a legend that Pythagoras' pupil Hippasus discovered this proof and was drowned at sea to keep it a secret.

[2] The current draft of this document is on the Web at
<http://www.cs.kun.nl/~freek/comparison/comparison.ps.gz>

There are several other comparisons between provers in the literature, but those generally compare only two systems, and often compare the systems for applications in computer science instead of for mathematics. For instance, there are comparisons between NuPRL and Nqthm [5], HOL and Isabelle [2], HOL and ALF [1], Coq and HOL [12], and HOL and PVS [11].

There also has been work done on how to embed the proofs of one system in another one. As an example, there is the work by Doug Howe and others on how to translate HOL proofs to classical NuPRL [16]. Surprisingly, mapping mathematics between systems is more difficult than one would expect.

1.4 Outline

In Section 2 we list the fifteen systems that are compared in this paper. We explain why we selected these systems and also why we did not select some other systems. In Section 3 we investigate various proof representations that are used by the fifteen systems. We give a classification of these representations into seven categories. Then we investigate the sizes of the irrationality of $\sqrt{2}$ proofs. Finally we compare the sizes of the libraries of the systems. In Section 4 we compare the logical foundations of the systems. We also look at their architecture according to the so-called 'de Bruijn criterion'. In Section 5 we compare the level of automation of the systems. We study whether they satisfy the 'Poincaré principle', whether they have an open architecture that allows user automation, and whether they come with strong built-in automation. Finally in Section 6 we put all systems in one diagram.

2 From Alfa to Ωmega: The Fifteen Provers of the World

The systems that are compared in this paper are listed in the following table:

1. **HOL**
   Web page: <http://www.cl.cam.ac.uk/Research/HVG/HOL/>
   Implementation language: ML
   Main person behind the system: Mike Gordon
   People who did $\sqrt{2} \notin \mathbb{Q}$: John Harrison, Konrad Slind

2. **Mizar**
   Web page: <http://mizar.org/>
   Implementation language: Pascal
   Main person behind the system: Andrzej Trybulec
   Person who did $\sqrt{2} \notin \mathbb{Q}$: Andrzej Trybulec

3. **PVS**
   Web page: <http://pvs.csl.sri.com/>
   Implementation languages: Lisp, ML
   Main people behind the system: John Rushby, Natarajan Shankar, Sam Owre
   People who did $\sqrt{2} \notin \mathbb{Q}$: Bart Jacobs, John Rushby

4. Coq

Web page: <http://pauillac.inria.fr/coq/>

Implementation language: ML

Main people behind the system: Gérard Huet, Thierry Coquand, Christine Paulin

Person who did $\sqrt{2} \notin \mathbb{Q}$: Laurent Théry

5. Otter/Ivy

Web pages: <http://www.mcs.anl.gov/AR/otter/> and <http://www-unix.mcs.anl.gov/~mccune/acl2/ivy/>

Implementation language: C

Main people behind the system: William McCune, Larry Wos, Olga Shumsky

People who did $\sqrt{2} \notin \mathbb{Q}$: Michael Beeson, William McCune

6. Isabelle/Isar

Web pages: <http://www.cl.cam.ac.uk/Research/HVG/Isabelle/> and <http://isabelle.in.tum.de/>

Implementation language: ML

Main people behind the system: Larry Paulson, Tobias Nipkow, Markus Wenzel

People who did $\sqrt{2} \notin \mathbb{Q}$: Markus Wenzel, Larry Paulson

7. Alfa/Agda

Web page: <http://www.cs.chalmers.se/~catarina/agda/> and <http://www.math.chalmers.se/~hallgren/Alfa/>

Implementation language: Haskell

Main people behind the system: Thierry Coquand, Catarina Coquand, Thomas Hallgren

Person who did $\sqrt{2} \notin \mathbb{Q}$: Thierry Coquand

8. ACL2

Web page: <http://www.cs.utexas.edu/users/moore/acl2/>

Implementation language: Lisp

Main person behind the system: J Strother Moore

Person who did $\sqrt{2} \notin \mathbb{Q}$: Ruben Gamboa

9. PhoX

Web page: <http://lama-d134.univ-savoie.fr/sitelama/Membres/pages_web/RAFFALLI/phox.html>

Implementation language: ML

Main person behind the system: Christophe Raffalli

People who did $\sqrt{2} \notin \mathbb{Q}$: Christophe Raffalli, Paule Rozière

10. IMPS

Web page: <http://imps.mcmaster.ca/>

Implementation language: Lisp

Main people behind the system: William Farmer, Joshua Guttman, Javier Thayer

Person who did $\sqrt{2} \notin \mathbb{Q}$: William Farmer

11. Metamath

Web page: <http://metamath.org/>
Implementation language: C
Main person behind the system: Norman Megill
Person who did $\sqrt{2} \notin \mathbb{Q}$: Norman Megill

12. Theorema

Web page: <http://www.theorema.org/>
Implementation language: Mathematica
Main person behind the system: Bruno Buchberger
People who did $\sqrt{2} \notin \mathbb{Q}$: Markus Rosenkranz, Tudor Jebelean, Bruno Buchberger

13. Lego

Web page: <http://www.dcs.ed.ac.uk/home/lego/>
Implementation language: ML
Main person behind the system: Randy Pollack
Person who did $\sqrt{2} \notin \mathbb{Q}$: Conor McBride

14. NuPRL

Web page: <http://www.cs.cornell.edu/Info/Projects/NuPrl/nuprl.html>
Implementation languages: ML, Lisp
Main person behind the system: Robert Constable
Person who did $\sqrt{2} \notin \mathbb{Q}$: Paul Jackson

15. Ωmega

Web page: <http://www.ags.uni-sb.de/~omega/>
Implementation language: Lisp
Main person behind the system: Jörg Siekmann
People who did $\sqrt{2} \notin \mathbb{Q}$: Christoph Benzmüller, Armin Fiedler, Andreas Meier, Martin Pollet

For most systems it is clear why they are in this list, but a few need explanation.

Otter is not designed for the development of a structured body of mathematics in the QED style, but instead is used in a 'one shot' way to solve logical puzzles. Also it is only one of the members (although the best known) of the large class of *first order theorem provers*. The reason that we still have included Otter in this list (and only Otter) is that Art Quaife has used Otter to develop a body of mathematics in Euclidean geometry and set theory [20]. Also Otter is the only program in the list that has been used for the solution of open mathematical problems, as listed in <http://www-unix.mcs.anl.gov/AR/new_results/>. A reason to single out Otter from the first order provers is that Otter has the Ivy program for separate checking of its proofs.

ACL2 is not primarily designed for mathematics. In particular it has a rather weak logic, without an explicit existential quantifier. However the Nqthm system (a predecessor of ACL2, which is very similar) has been used to formalize significant theorems like Gödels first incompleteness theorem [21]. Also Nqthm was the system that the authors of the QED manifesto had in mind.

The Metamath system [14] maybe should not be counted as an 'industrial strength' system: it only has one user. However, the system is beautifully executed and differs in many respects from the other systems. For one thing it is very fast: it can check its full (non-trivial) library in only a few seconds. Also it really makes the logical structure of the mathematics completely transparent.

Some of the systems in the list have predecessors or (recent) successors:

ALF, Half	$\rightarrow$	**Agda**
Nqthm	$\rightarrow$	**ACL2**
Imps	$\rightarrow$	MathScheme
Lego	$\rightarrow$	Oleg, Plastic
NuPRL	$\rightarrow$	MetaPRL

We did not include any of those. We expect systems from the same origin to be reasonably close to each other.

Some recent provers, like the KIV system and the 'B method', have been especially designed for verification of hardware and software. These systems can also be used for the formalization of mathematics, but we have not included them in our comparison. Other systems, like the Twelf system and the Typelab system, are more for formalizing logic than for mathematics. After some discussion with the authors of these systems we decided to omit these as well. Finally there is the TPS system which is similar to Otter but for higher order logic. However, it misses what makes Otter interesting for this comparison, so it was left out too.

3 Files Containing Mathematics

3.1 Seven Ways to Represent Mathematics

When we were editing the fifteen proofs of the irrationality of $\sqrt{2}$ for presentation on paper, it turned out that often it was difficult to present the proofs in such a way that it was clear what was going on. For various systems we needed to show the proof in multiple representations. Some reflection showed that these representations could be divided into seven groups:

	HOL	Mizar	PVS	Coq	Otter/Ivy	Isabelle/Isar	Alfa/Agda	ACL2	PhoX	IMPS	Metamath	Theorema	Lego	NuPRL	Ωmega
definitions & statements of lemmas	i	i	i	i	i	i	i	i	i	i	i	x	i	i	i
proof scripts	=	=	i	=		=	=		=	=	=		=	i	i
trace of interactive session	o		o	o	o	o		o			o	=	o	o	o
representation with symbols							=	=		x	o	=		=	o
representation in natural language	=							=	=			=			o
stored formalization state				o										o	o
λ-term or other 'proof object'				o	o	o	=						o		=

In this table an '*i*' means 'input file' that has been made by the user. A '*o*' means 'output file' that has been generated by the system. An '*x*' means that it is a mixture of input from the user and output from the system. An '=' sign means that this is part of the same file as the item above it in the same column. For instance, in the Mizar system the statement of the lemmas, the 'proof scripts' that the user enters, and the 'natural language' representation, all are in one file (the .miz file) which is written by the user.

In this table we only included files that can be displayed on paper. For some systems the stored formalization state is not represented in this table because it is a binary file which is not humanly readable.

3.2   Comparing the Sizes of the Input Files

We now compare the sizes of the formalization of the irrationality of $\sqrt{2}$ in the fifteen systems. This not only compares the systems but also the styles of the people who did the formalization and the complexity of the proof that they selected for formalization. Therefore it only gives a rough indication of the 'compactness' of the systems.

Also, comparing systems based on a single proof problem is statistically not meaningful. Having the same proof in fifteen systems does show the proof styles of the systems surprisingly well, but it does not say much about the quality of the provers. Still, we will list the sizes of the files.

Here are the precise statements that were proved (or, in the case of Otter, disproved) in the fifteen systems:

HOL	`~rational(sqrt(&2))`
Mizar	`sqrt 2 is irrational`
PVS	`NOT Rational?(sqrt(2))`
Coq	`(irrational (sqrt (S (S 0))))`
Otter	`m(a,a) = m(2,m(b,b))`
Isabelle/Isar	$sqrt\ (real\ (2::nat)) \notin \mathbb{Q}$
Alfa/Agda	$prime\ p \rightarrow noether\ A\ (multiple\ p) \rightarrow isNotSquare\ p$
ACL2	`(implies (equal (* x x) 2)`
	`           (and (realp x)`
	`                (not (rationalp x))))`
PhoX	`/\m,n : N (m^ N2 = N2 * n^ N2 -> m = N0 & n = N0)`
IMPS	`not #(sqrt(2),qq)`
Metamath	`$p |- ( sqr ` 2_10) e/ QQ`
Theorema	$\neg\mathrm{rat}\left[\sqrt{p}\right]$
Lego	`{b|nat}{a|nat}`
	`(Eq (times two (times a a)) (times b b))->`
	`(Eq a zero /\ Eq b zero)`
NuPRL	$\neg(\exists u:\mathbb{Q}.\ u *_q u = 2 / 1)$
Ωmega	`(not (rat (sqrt 2)))`

Some people proved the statement of the irrationality in the real numbers:

$$\sqrt{2} \notin \mathbb{Q}$$

Others did not have a library of real numbers (or did not want to use it) and only proved a statement about natural numbers:

$$m^2 = 2n^2 \iff m = n = 0$$

The Agda proof by Thierry Coquand proves something still more basic. It does not talk about the number two in the natural numbers, but instead about any element in a commutative monoid that satisfies some conditions. The Otter proof has a similar structure.

Most people proved the irrationality of the square root of two, but some only proved the irrationality of an arbitrary prime number (where 'prime' means that the number divides a product if and only if it divides one of the factors). This might sound stronger, but Conor McBride noted that it is the other way around: it is also needs some non-trivial work to prove that two is prime.

Most people proved the statement using the library of their system, but not all systems have a library. In that case some lemmas were proved from statements that were taken as axioms. The Ωmega system does have a standard library, but not all statements in this library have been proved using the system. For the irrationality of $\sqrt{2}$ four lemmas were added to this library, but only one was proved. The Agda system does not have a library, but the formalization did not use any unproved statements. It defined everything that was used, including the logic.

The IMPS proof is by far the largest in the collection. It could have been quite a bit shorter, but William Farmer chose to first prove the more general statement that the square root of any non-square number is irrational.

	lines			fragment
Otter	17	monoid	prime	one lemma
HOL	29	$\mathbb{R}$	2	from library
Ωmega	38*	$\mathbb{R}$	2	from library, one lemma out of four
Theorema	39*	$\mathbb{R}$	prime	two lemmas
Mizar	44	$\mathbb{R}$	2	from library
NuPRL	54*	$\mathbb{N}$	2	from library
Coq	68	$\mathbb{R}$	2	from library
PVS	77	$\mathbb{R}$	2	from library
Metamath	81	$\mathbb{R}$	2	from library
Isabelle	114	$\mathbb{R}$	2	from library
PhoX	151	$\mathbb{N}$	2	from library
ACL2	206	$\mathbb{R}$	2	from library
Agda	230	monoid	prime	stand alone
Lego	261	$\mathbb{N}$	2	from library
IMPS	663	$\mathbb{R}$	2	from library

For this table we only counted non-blank non-comment lines.[3] The line counts that have been marked with an asterisk do not refer to a specific file, but instead are combined counts of relevant parts of files. If we restrict ourselves to formalizations that prove the full statement about the number two in the real numbers without omitting anything, then the three shortest proofs are the HOL, Mizar and Coq proofs.

3.3 The Library

In practice for serious formalization of mathematics a good library is more important than a user friendly system. The Mizar systems has the largest library by far. It proves over 32 thousand lemmas, taking 50 megabytes or 1.4 million lines.

The following systems currently have a large *mathematical* library:

	HOL	Mizar	PVS	Coq	Otter/Ivy	Isabelle/Isar	Alfa/Agda	ACL2	PhoX	IMPS	Metamath	Theorema	Lego	NuPRL	Ωmega
large mathematical library	•	•	•	•		•					•			•	

4 Differences in Logical Strength

4.1 Logics and Type Systems

The systems vary in underlying logic and type system. Here is a table that shows the different logics of the fifteen systems:

	HOL	Mizar	PVS	Coq	Otter/Ivy	Isabelle/Isar	Alfa/Agda	ACL2	PhoX	IMPS	Metamath	Theorema	Lego	NuPRL	Ωmega
primitive recursive arithmetic								•							
first order logic					•										
higher order logic	•		•			•			•	•		•			•
first order set theory		•									•				
higher order type theory				•			•						•	•	
classical logic	•	•	•		•	•		•	•	•	•	•			•
constructive logic				•			•						•	•	
quantum logic											•				
fixed logic	•	•	•	•	•		•	•	•	•	•		•	•	•
logical framework						•						•			

---

[3] The relevant files are on the Web in
<http://www.cs.kun.nl/~freek/comparison/comparison.tar.gz>

A logical framework does not just support some given logics, but instead the user is able to define logics of his own. In the case of a logical framework the first two sections of the table indicate the most commonly used logics of the system.

Coq and NuPRL are implementations of variants of intuitionistic type theory. The 'classical variants' of these systems are equiconsistent with ZFC set theory with countably many inaccessible cardinals.[4] In particular, they are quite a bit stronger than the higher order logics of systems like HOL.

The Mizar system is logically even stronger than this, because it has arbitrarily large inaccessibles. However, Mizar is clearly a first order system.[5] The Coq and NuPRL systems are higher order systems.[6]

Here is a table that shows the type systems of the fifteen systems (a system is only considered typed when the types are first class objects that occur in variable declarations and quantifiers):

	HOL	Mizar	PVS	Coq	Otter/Ivy	Isabelle/Isar	Alfa/Agda	ACL2	PhoX	IMPS	Metamath	Theorema	Lego	NuPRL	Ωmega
untyped					•			•			•	•			
decidable non-dependent types	•	•				•			•	•					•
decidable dependent types				•			•						•		
undecidable dependent types			•											•	

4.2 The de Bruijn Criterion

The de Bruijn criterion states that the correctness of the mathematics in the system should be guaranteed by a *small* checker. Architecturally this generally means that there is a 'proof kernel' that all the mathematics is filtered through. In the HOL Light variant of the HOL system this kernel is extremely small: it consists of only 285 lines of ML. The NuPRL system is just over the border of this criterion. It has a proof checking kernel but this kernel is not small.

---

[4] In the NuPRL system the classical variant is officially supported. In the Coq system the equiconsistency actually seems to break down, because the impredicativity of Coq is inconsistent with classical mathematics.

[5] Steps from Mizar proofs correspond directly to first order problems [9]. As part of his PhD research, Josef Urban from the Charles University in Prague is developing software that can export any Mizar step in TPTP format.

[6] The NuPRL type theory is predicative, which means that it does not have just one type for propositions, but one for each type universe. However, in practice NuPRL is higher order logic: it can abstract and quantify over arbitrary higher order types.

	HOL	Mizar	PVS	Coq	Otter/Ivy	Isabelle/Isar	Alfa/Agda	ACL2	PhoX	IMPS	Metamath	Theorema	Lego	NuPRL	Ωmega
de Bruijn criterion	•			•	•	•	•		•		•		•		•

The Otter system does not have a proof checking kernel built into the system. However, there is the Ivy system that can export Otter proofs in a form that can be checked by a very small ACL2 program.

5   Proof Checking or Theorem Proving

5.1   Interaction Styles

Among the systems one finds three different interaction styles. First, there are the systems in which the user writes the text of the proof and the system checks the correctness afterwards. Second, there are the 'proof assistants' which keep a proof state for the user. The user then modifies this proof state through the application of so-called 'tactics'. Third, there are the automated theorem provers which automatically prove lemmas that the user states. The involvement of the user then only consists of the selection of the lemmas (as 'stepping stones' towards the final result) and the selection of parameters for the prover.

	HOL	Mizar	PVS	Coq	Otter/Ivy	Isabelle/Isar	Alfa/Agda	ACL2	PhoX	IMPS	Metamath	Theorema	Lego	NuPRL	Ωmega
proof checking				•			•								
goal transformation through tactics	•		•	•		•					•	•	•	•	•
automated theorem proving					•			•			•				

Note that tactic-based provers can still have powerful automation. For instance the PVS system has powerful decision procedures.

5.2   The Poincaré Principle and Automation

An important aspect of a mathematical system is automation of trivial tasks. In particular a user should not need to spell out calculations in detail. A system that can prove the correctness of calculations automatically is said to satisfy the Poincaré principle [3],[7] because in [19] Henri Poincaré wrote about showing the correctness of the calculation $2 + 2 = 4$:

---

[7] Henk Barendregt links the Poincaré principle to reduction in type theory, but this is only partially correct. The Agda system has $\beta\delta\iota$-reduction but it cannot use it for

'Ce n'est pas une démonstration proprement dite, [...] c'est une vérification'. [...] La vérification diffère précisément de la véritable démonstration, parce qu'elle est purement analytique et parce qu'elle est stérile.

	HOL	Mizar	PVS	Coq	Otter/Ivy	Isabelle/Isar	Alfa/Agda	ACL2	PhoX	IMPS	Metamath	Theorema	Lego	NuPRL	Ωmega
Poincaré principle	•	•	•	•	•		•	•		•		•	•	•	•
user automation	•		•	•		•		•						•	•
powerful built-in automation	•	•		•	•		•		•		•		•		•

An important aspect of a prover is whether it has an 'open' architecture, i.e., whether the user can write programs to solve proof problems algorithmically. Most provers allow *some* automation like this, but in the table we indicate whether this programmability is on the level of the implementation of the system.

Finally there is the aspect whether a prover already has strong automation built-in.[8] Examples of such automation are decision procedures for algebraic problems, proof search procedures, and automation of induction.

6   A Rather Subjective Two-Dimensional Diagram

We compiled some of the information about the systems into a diagram (like the Herzsprung-Russell diagram classifying stars in astronomy). On the horizontal axis the diagram shows how 'mathematical' the logic of the system is. On the vertical axis it shows how much automation the system offers. The sizes of the circles correspond to the sizes of the respective mathematical libraries.

The positions of the systems in this diagram are rather subjective. Most circles can be argued around quite a bit. To make the diagram somewhat objective we have 'scored' various aspects of the systems. This made the diagram turn out to be surprising to us: we had expected the Theorema and IMPS systems to score more mathematical than they do in this diagram, and the Otter system to score higher on the automation axis.

---

reflection as it lacks the automation to lift the expressions to the syntactical level, so we claim that Agda does not satisfy the Poincaré principle. On the other hand HOL does not have reduction in its logic, but it is easy to program HOL 'conversions' to prove calculations automatically.

[8] We only considered a tactic-based prover to have 'powerful built-in automation' if it has a tactic that is a full first order prover, like Isabelle's *blast*.

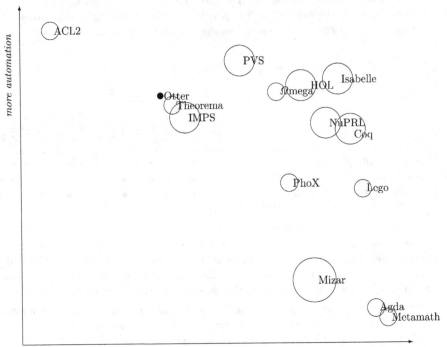

Items that added to the score for 'more mathematical' were:

- more powerful logic
- logical framework
- dependent types
- de Bruijn criterion

while items that added to the score for 'more automation' were:

- more automated interaction style
- Poincaré principle
- user automation
- powerful built-in automation

7   Future Work

The comparison in this paper does not focus in detail on the automation of the systems. It is worthwhile to investigate the various kinds of automation of these systems in more detail, and in particular to investigate their algebraic decision procedures.

References

1. S. Agerholm, I. Beylin, and P. Dybjer. A Comparison of HOL and ALF Formalizations of a Categorical Coherence Theorem. In *TPHOLs'96*, volume 1125 of *LNCS*, pages 17–32. Springer-Verlag, 1996.
2. S. Agerholm and M.J.C. Gordon. Experiments with ZF Set Theory in HOL and Isabelle. In *8th International Workshop on Higher Order Logic Theorem Proving and its Applications*, volume 971, pages 32–45. Springer-Verlag, 1995.
3. Henk Barendregt. The impact of the lambda calculus. *Bulletin of Symbolic Logic*, 3(2), 1997.
4. Henk Barendregt and Herman Geuvers. Proof-Assistants Using Dependent Type Systems. In Alan Robinson and Andrei Voronkov, editors, *Handbook of Automated Reasoning*. Elsevier Science Publishers B.V., 2001.
5. D. Basin and M. Kaufmann. The Boyer-Moore Prover and NuPRL: An experimental comparison. In *Proceedings of the First Workshop on 'Logical Frameworks'*, *Antibes, France*, pages 89–119. Cambridge University Press, 1991.
6. R. Boyer et al. The QED Manifesto. In A. Bundy, editor, *Automated Deduction – CADE 12*, volume 814 of *LNAI*, pages 238–251. Springer-Verlag, 1994. <http://www.cs.kun.nl/~freek/qed/qed.ps.gz>.
7. B. Buchberger, T. Jebelean, F. Kriftner, M. Marin, and D. Vasaru. An Overview on the Theorema project. In W. Kuechlin, editor, *Proceedings of ISSAC'97 (International Symposium on Symbolic and Algebraic Computation)*, Maui, Hawaii, 1997. ACM Press.
8. Robert L. Constable, Stuart F. Allen, H.M. Bromley, W.R. Cleaveland, J.F. Cremer, R.W. Harper, Douglas J. Howe, T.B. Knoblock, N.P. Mendler, P. Panangaden, James T. Sasaki, and Scott F. Smith. *Implementing Mathematics with the Nuprl Development System*. Prentice-Hall, NJ, 1986.
9. Ingo Dahn and Christoph Wernhard. First Order Proof Problems Extracted from an Article in the MIZAR Mathematical Library. In *Proceedings of the International Workshop on First order Theorem Proving*, number 97-50 in RISC-Linz Report Series, pages 58–62, Linz, 1997. Johannes Kepler Universität.
10. G.P. Goold, editor. *Selections illustrating the history of Greek mathematics, with an English translation by Ivor Thomas*. Harvard University Press, London, 1939.
11. David Griffioen and Marieke Huisman. A comparison of PVS and Isabelle/HOL. In Jim Grundy and Malcolm Newey, editors, *Theorem Proving in Higher Order Logics: 11th International Conference, TPHOLs'98*, volume 1479 of *LNCS*, pages 123–142. Springer-Verlag, 1998.
12. L. Jakubiec, S. Coupet-Grimal, and P. Curzon. A Comparison of the Coq and HOL Proof Systems for Specifying Hardware. In E. Gunter and A. Felty, editors, *International Conference on Theorem Proving in Higher Order Logics: B-Track*, pages 63–78, 1997.
13. Matt Kaufmann, Panagiotis Manolios, and J. Strother Moore. *Computer-Aided Reasoning: An Approach*. Kluwer Academic Publishers, Boston, 2000.
14. Norman D. Megill. Metamath, A Computer Language for Pure Mathematics. <http://metamath.org/>, 1997.
15. M. Muzalewski. *An Outline of PC Mizar*. Fondation Philippe le Hodey, Brussels, 1993. <http://www.cs.kun.nl/~freek/mizar/mizarmanual.ps.gz>.
16. P. Naumov, M.-O. Stehr, and J. Meseguer. The HOL/NuPRL Proof Translator: A Practical Approach to Formal Interoperability. In R.J. Boulton and P.B. Jackson, editors, *The 14th International Conference on Theorem Proving in Higher Order Logics*, volume 2152 of *LNCS*, pages 329–345. Springer-Verlag, 2001.

17. R.P. Nederpelt, J.H. Geuvers, and R.C. de Vrijer. *Selected Papers on Automath*, volume 133 of *Studies in Logic and the Foundations of Mathematics*. Elsevier Science, Amsterdam, 1994.
18. S. Owre, J. Rushby, and N. Shankar. PVS: A prototype verification system. In D. Kapur, editor, *11th International Conference on Automated Deduction (CADE)*, volume 607 of *LNAI*, pages 748–752, Berlin, Heidelberg, New York, 1992. Springer-Verlag.
19. Henri Poincaré. *La Science et l'Hypothèse*. Flammarion, Paris, 1902.
20. A. Quaife. *Automated Development of Fundamental Mathematical Theories*. Kluwer Academic, 1992.
21. N. Shankar. *Metamathematics, Machines and Gödel's Proof*. Number 38 in Cambridge Tracts in Theoretical Computer Science. Cambridge University Press, 1994.
22. Otto Töplitz. *The Calculus: A Genetic Approach*. University of Chicago Press, Chicago, 1963. Translated by Luise Lange.
23. F. Wiedijk. Mizar: An Impression. `<http://www.cs.kun.nl/~freek/mizar/mizar intro.ps.gz>`, 1999.

Translating Mizar for First Order Theorem Provers

Josef Urban

Dept. of Theoretical Computer Science
Charles University
Malostranske nam. 25, Praha, Czech Republic
urban@kti.ms.mff.cuni.cz

Abstract. The constructor system of the Mizar proof checking system is explained here on examples from Mizar articles, and its translation to untyped first-order syntax is described and discussed. This makes the currently largest library of formalized mathematics available to first-order theorem provers.[1]

1 Introduction, Previous Work

Mizar [Rudnicky 92] is a system for computer checked mathematics. In more detail, Mizar is associated with several things:

- The Mizar language ... this is the language in which Mizar articles must be written, so that they can be checked by computer.
- The Mizar Mathematical Library (MML). This is the growing (now about 700) collection of Mizar articles that have already been written and computer checked and the notation, definitions, theorems and other Mizar constructs created in them can be used for writing new Mizar articles. Various presentations of the MML exist today: Formalized Mathematics, online html-ized abstracts, Mizar Encyclopedia.
- The Mizar checker and other software utilities for working with articles and MML.
- The Mizar project headed by A.Trybulec, taking care of the things named above as well as other things related to Mizar.

Several introductions to the Mizar language as well as to the practical aspects of writing Mizar articles exist today e.g. [Bonarska 90, Muzalewski 93, Rudnicky 92] or [Wiedijk 99]. It is recommended to have a look at one of them, but we will try here to explain the features of Mizar, relevant for the first order translation.

The largest previous attempt at first order translation of the MML was done by the ILF group [Dahn 97]. However, the work remained unfinished and only

---

[1] A more detailed version of this article will be a part of author's PhD thesis.

A. Asperti, B. Buchberger, J.H. Davenport (Eds.): MKM 2003, LNCS 2594, pp. 203–215, 2003.

several basic articles were translated. The translation we present is done uniformly for the whole first order part of MML[2], and tested even for very advanced MML articles.

2   General Descriptions

2.1   The Mizar Language

The detailed specification of the Mizar language is given in [Syntax] and notions defined there will be written in italics here. We recommend to have a look at a sample Mizar article (e.g. article about real numbers [REAL_1]), to see examples of the abstract syntax. The logic (especially constructor and type system) of Mizar has not been formalized yet (strictly speaking, it is formalized in the Mizar sources). So our approach is to follow the [Syntax], explaining what the language constructs mean in the Mizar implementation, and map them into untyped first order syntax. The most practical "proof" of the correspondence of the translation with the Mizar sources, is then the correspondence of the results of the Mizar checker with the results of theorem provers run on translated tasks.

The main parts of a Mizar article are definitions (*Definitional-Items*,(*Functor, Predicate, Mode, Structure and Attribute-Definition*)), *Theorems* and *Schemes*. Attributes are predicates handled speciallly by the system, according to the rules defined by the user in *Cluster-Registrations*. All Mizar terms are typed. There is a largest (default) type called "set" or "Any". All other types have one or more mother types. Types of variables are given either in global *Reservations* or local *Loci-Declarations* or inside quantified formulas. Types of other terms are computed according to *Functor-Definitions*. Types can have arguments (be parameterized) in Mizar, e.g. "Element of X" or "Function of NAT, REAL" are legal types with one or two arguments, respectively. Type translation is the largest part of the first order translation. There are two possible basic approaches to type translation[3](we use the second approach, which is better for SPASS).

- Types can be thought of as set-theoretic classes (e.g. type "set" being the universal class, type "Element of X" being the set of all elements of X, type "Integer" being the set of all integers, etc.).
- Types can be thought of as predicates .... thus "set(X)" is true for any X, "Integer(X)" is true iff X is integer, and "Element(X,Y)" is true iff $X \in Y$

2.2   Syntactic Levels

The Mizar language handles a very large database of articles about different parts of mathematics and this necessarily leads to notation conflicts. The solution to such conflicts is introduction of two syntactic levels of the language: the **pattern** level and the **constructor** level.

---

[2] For the precise meaning of this, see section Unimplemented Features of Mizar.

[3] [Dahn 98] suggests another, based on inclusion operations ([Goguen 92]), but the system of attributes (and thus vast multisortedness of MML terms) makes it probably less efficient than the standard translations.

Constructors are the real unambiguous functions, predicates, types and attributes to which the patterns are translated before any proof checking takes place. Patterns are mapped to the constructors; they accommodate the need for having different symbols for the same constructor or vice versa (same symbol for different constructors). The process of mapping patterns to constructors is done separately for each Mizar article depending on its *Environment-Declarations* and is usually quite nontrivial.

Example 1. binary symbols "in" and "<'" define different patterns, but when dealing with ordinals, they are mapped to the same constructor, i.e. "A in B" and "A <' B" cannot be distinguished when translated to the constructor level.

Definitions can influence both the pattern and the constructor level. Typically, a definition causes a new pattern and also a new constructor to be defined, but in many cases there is no effect on the constructor level. For the purpose of translating Mizar articles into first order syntax, only the constructor level is important, the patterns can be thought of as a "syntactic sugar" added on top of the constructors.

2.3   First Order Formats, Outline of the Translation

Several kinds of first order syntax are used today for first-order provers, e.g. TPTP format [Suttner and Sutcliffe 98], DFG format [Hahnle et all 96], Otter format [McCune 94] and others. We chose direct translation into DFG format, since our immediate purpose was to experiment with the SPASS prover [Weidenbach et all 99]. More generic approach is certainly desirable, however, at a first glance, several such approaches come into mind (e.g. ILF-like approach [Dahn 97], TPTP-like approach, or even adding direct Mizar support for other formats than just DFG), so we postponed such decisions for later time. Note that the dfg2tptp tool can be used for translation to TPTP and thus (using TPTP tools) to virtually any other syntax.

So the general approach to translation is following:

- We use parts of Mizar (Parser, Analyzer) which translate the Mizar articles to the constructor level, where our first order translation starts.
- We give absolute (context independent) names to all constructors[4].
- Definitions of constructors usually translate to several first order formulas, since we have to translate both the type hierarchy information given in the definitions and the actual *Definiens*.
- Sometimes also additional properties (e.g. commutativity, transitivity, etc.) of the defined constructors are stated inside definitions. They are also translated into corresponding first order formulas.
- All formulas are relativized with respect to the typed variables occurring in them (using the above mentioned predicate translation of types). So e.g. universally quantified Mizar formula

---

[4] The naming scheme we use in what follows has obviously no importance for theorem provers, however, it has become standard in various Mizar presentations, allowing their interoperability.

"for x being Real holds x-x = 0" translates to first order formula "Real(x) implies x-x = 0".

Remark: The actual DFG syntax translation of the above mentioned Mizar formula is:
`forall([B1], implies(v1_arytm(B1), equal(k3_real_1(B1,B1),0)))`.
However, for explanation purposes, we use here rather the user-defined symbols instead of the absolute constructor names, and also for the sake of better readability, we do not strictly adhere to the DFG syntax of formulas.

Next we explain in more detail the translation of various Mizar definitions.

3   Translation of Mizar Definitions

3.1   Mode Definitions

Types in Mizar can be defined using either *Mode-Definitions* or *Structure-Definitions*. We deal with modes first. Since structures are more complicated, we postpone them after the explanation of functions.

As already stated, the translation of the largest (default) type "set" is a predicate that always holds true ("set(X)" is true for all terms X). So omitting relativization by this predicate is logically correct, and we do it for the sake of better readability of translated formulas. For all other modes, the syntactic structure of *Mode-Definitions* is:

mode *Mode-Pattern* ([*Specification*] [means *Definiens*]
Correctness-Conditions ; | is *Type-Expression* ;) { *Mode-Synonym* } .

Example 2.
```
definition let X;
 mode Element of X means
:Def2: it in X if X is non empty
    otherwise it is empty;
 existence by BOOLE:def 1;
 consistency;
end;
```

This is a definition of the mode "Element of X" from the article subset_1.miz. The *Loci-Declaration* "let X;" declares the variables occurring in the definition. This definition omits the optional *Specification* part. If the *Specification* is present in the definition, it gives a mother type for the newly defined mode. If not, the largest type "set" is used as the mother type.

The *Definiens* after the keyword "means" consists of an optional *Label-Identifier* and either a simple sentence or several sentences separated by the keywords "if" and "otherwise". The latter (*Conditional-Definiens*) is used for "per cases" definitions.

In our example, the *Conditional-Definiens* distinguishes the case when the mode argument X is non empty from the case when the mode argument X is empty. The sentence for the first case states, that for any "it" (special variable used in definitions) being of the type "Element of X" means "it in X". The second case sentence says, that in the degenerate case when X is empty, being of the type "Element of X" means to be empty too[5].

The keywords "existence" and "consistency" in the example introduce *Correctness-Conditions*. The "existence" condition states that the extension of the defined type is non empty. The "consistency" condition states that the disjunction of all the cases into which the definition was split, is true. The absolute name given to the mode constructor created in this definition is "m1_subset_1", which means that it is the first mode constructor in article subset_1,

We translate types as predicates, so n-ary types will become n+1-ary predicates, and e.g. the Mizar type qualification "X is Element of Y" will be translated as "m1_subset_1(X,Y)".

Next we want to encode the part of the type hierarchy created by this definition. As noted above, the mother type of "Element of X" is the largest type "set". The translation would be following:

‘‘forall([X,Y], m1_subset_1(Y,X) implies set(Y)).’’

However, as already mentioned, "set(Y)" is always true and we chose not to translate it, so in such cases (i.e. when the mother type is "set"), we do not create the type hierarchy translation.

The definiens translates to the following formula:

‘‘m1_subset_1(Y,X) iff ((not(empty(X)) and in(Y,X))
or (empty(X) and empty(Y)))’’

(instead of "empty" and "in" their absolute names would be used).
Finally, we translate the existence condition:

‘‘forall([X], exists([Y], m1_subset_1(Y,X))).’’

3.2   Predicate and Attribute Definitions

In comparison to first order predicates, Mizar predicates can have several predefined properties (e.g. reflexivity, symmetry, etc.), and they also define the type of their arguments. *Predicate-Definition* has the following syntactic structure:
pred *Predicate-Pattern* [means *Definiens*] ; *Correctness-Conditions*
{ *Predicate-Property* } { *Predicate-Synonym* } .

```
e.g.:
 definition let X,Y;
   pred X c= Y means
:: TARSKI:def 3
   x in X implies x in Y;
   reflexivity;
 end;
```

This is a definition of the subset relation from article tarski. Again first comes a *Loci-Declaration* "let X,Y;", after that the *Predicate-Pattern* "X c=

---

[5] If the reader is curious about why the degenerate case is handled this way, the answer is, that simply in many cases it turned out to be advantageous.

Y", then the *Definiens* "x in X implies x in Y;" and finally one *Predicate-Property* is stated: "reflexivity;". *Correctness-Conditions* occur only in re-definitions of predicates, redefinitions will be discussed later.

Again, this definition creates both a pattern and a constructor, the standard symbol for predicate constructors is "r", so the absolute name would be "r1_tarski" here.

The *Definiens* formula is simply translated as equivalence:
''r1_tarski(X,Y) iff (x in X implies x in Y)''

Predicate-Property can be:
symmetry connectedness reflexivity irreflexivity
So we add the corresponding theorem for such properties, here it is:
''r1_tarski(X,X).''

Attributes are very similar to predicates, so due to space constraints, their explanation is omitted here.

The *Cluster-Registrations*, which define rules of attribute handling will be explained later.

A possibly negated attribute is called *Adjective* in Mizar. A finite set of *Adjectives* is called *Adjective-Cluster* (or just cluster) in Mizar. Clusters can be used as prefixes to types, e.g. "finite non empty set" uses the *Adjectives* "finite" and "non empty" as a prefix to the type "set". Such types with non empty *Adjective-Cluster* are translated as a conjunction of all the predicates corresponding to the attributes and the predicate corresponding to the underlying type, so here it is:
"finite(X) and not(empty(X)) and set(X)" for a variable X of the type "finite non empty set".

Again, the "set" can be pruned, so the translations is just ''finite(X) and not(empty(X))''.

3.3   Functor Definitions

Functions in Mizar have to define the types of their arguments and the type of their result. Several *Functor-Properties* (e.g. "commutativity") can be associated with them.

The syntax of *Functor-Definition* is:
func *Functor-Pattern* [*Specification*] [(means | equals) *Definiens*] ; *Correctness-Conditions* { *Functor-Property* } { *Functor-Synonym* } .

Example 3.
definition let x be real number;
   func -x -> real number means :Def1: x + it = 0;
   existence by AXIOMS:19;
   uniqueness by Th10;

The initial *Loci-Declaration* declares a variable x for the type "number" with *Adjective* "real". Type "number" is now defined just as a convenient synonym for the largest type "set" (in article arytm), "real" is an attribute defined also in article arytm. The *Functor-Pattern* is "-x" here, and the *Specification* "->

real number" defines the result type of the function. The *Definiens* ":Def1: x + it = 0;" is the definitional formula of the defined function.

The *Correctness-Condition* "existence" states, that there exists an object (of the desired type "real number") conforming to the *Definiens*. "uniqueness" says that such object is unique (proved by previous theorem Th10). Before accepting a new *Functor-Definition*, the system always checks these two conditions.

This definition creates one new pattern and one new constructor. The standard Mizar symbol for functor constructors is "k" so the constructor's absolute name is "k1_real_1" (first functor in article real_1). We need to translate the typing, the *Definiens* and possibly the *Functor-Properties* (none in this case).

The typing translates to:

"'real number'(X) implies 'real number'(k1_real_1(X))."

Following the note above about *Adjective-Clusters*, 'real number'(X) translates to "(real(X) and number(X))" and since "number" is just a synonym for "set" it is pruned to just "real(X)". Since "real"'s absolute name is "v1_arytm", the exact translation is:

"forall([X], implies(v1_arytm(X), v1_arytm(k1_real_1(X))))".

The *Definiens* is translated by first instantiating the special variable "it" with the *Functor-Pattern*, yielding:

"x + (-x) = 0;" and translating the result formula, which gives:

"equal(+(x,k1_real_1(x)),0)".

After relativization and replacing "+" with its absolute name "k3_arytm", we get:

"forall([X], implies(v1_arytm(X), equal(k3_arytm(X,k1_real_1(X)),0)))"

3.4 Structure Definitions

Structures correspond (to some extent) to the "product" types of various other languages. The syntax of *Structure-Definition* is:

struct [(*Ancestors*)] *Structure-Symbol* [over *Loci*] (# *Fields* #) ;

Example 4.

struct(1-sorted) TopStruct (# carrier -> set,
                                topology -> Subset-Family of the carrier #);

This is definition of the structure TopStruct from the article about Topological Spaces "pre_topc"[6].

Structures can define their ancestor structures. All *Fields* of an ancestor must also be *Fields* of the defined structure. In our example, there is given one ancestor "1-sorted". *Ancestors* of structures are treated in a way similar to the treatmnt of mother types of modes by the system, i.e. they also define the type hierarchy.

The *Structure-Symbol* is "TopStruct" in our example. The optional syntax "[over *Loci*]" is not used there. If used, it parameterizes the structure in the

---

[6] This is not yet a topological space, just the underlying structure.

same way as modes can be parameterized, i.e. structures can also have any number of arguments. Finally, the structure defines two *Fields*:

"carrier" of the type "set", and

"topology" of the type "Subset-Family of the carrier".

We will now explain the effects of the *Structure-Definition* on the constructor level. Any structure defines a new type constructor. Such constructors are also called "aggregate types", and the standard Mizar symbol for them is "l". So the absolute name of the aggregate type in our example is "l1_pre_topc". Such types behave in exactly the same way as the mode types, i.e. variables can be reserved for them, *Adjective-Clusters* can be added to them as a prefix, etc. *Ancestors* become the mother types of the new aggregate type, so in our example, there is one mother type "1-sorted" (its absolute name is "l1_struct_0").

The *Fields* of a structure give rise to the so called "selector constructors". The standard Mizar symbol for selectors is "u". In our case, the structure contains two selectors, but the selector "carrier" is inherited from the ancestor "1-sorted" (l1_struct_0), where it was defined for the first time. So its absolute name is "u1_struct_0". The selector "topology" is defined for the first time in our example, so its absolute name is "u1_pre_topc".

The meaning of a selector is a function operating on the aggregate type. So e.g. "carrier" takes arguments of the type "1-sorted" (or any of its specializations, like TopStruct) and its result type is "set".

Finally, we need something to create the desired structure, when the *Fields* are given. This is solved in Mizar by creating "aggregate functor" for the structure. The standard Mizar symbol for aggregate functors is "g", so for TopStruct, its absolute name is "g1_pre_topc". The symbol for the aggregate functor is in Mizar the same as the symbol for the aggregate type, i.e. TopStruct. Given properly typed arguments, e.g. by following global reservations:

"reserve X for set;"

"reserve Y for Subset-Family of X;"

the Mizar term "TopStruct(#X,Y#)" means applying the aggregate functor "TopStruct" ("g1_pre_topc") to arguments X and Y, yielding an object of the aggregate type "TopStruct" ("l1_pre_topc").

Obviously, it holds then that

"X = the carrier of TopStruct(#X,Y#)" and

"Y = the topology of TopStruct(#X,Y#)".

Aggregate types are translated again as predicates, selectors and aggregate functors are translated as functions.

Axiom stating mutual interdependence of these constructors has to be added, this is done by stating that the structures are freely generated by its *Fields*, using the aggregate functor:

"TopStruct(X) implies X = TopStruct(carrier(X),topology(X));",

in absolute notation:

"forall([X], implies(l1_pre_topc(X),

equal(X,g1_pre_topc(u1_struct_0(X),u1_pre_topc(X)))))".

3.5   Cluster Registrations

We have already mentioned, that *Cluster-Registrations* define for the system the special rules of attribute handling. They either state non emptiness of attributed types (*Existential-Registration*) or define attribute propagation (*Conditional-Registration, Functorial-Registration*).

Existential-Registration. The syntax of *Existential-Registration* is:
cluster *Adjective-Cluster Type-Expression* ; *Correctness-Conditions.*

Existential-Registrations assert the existence (non emptiness of the extension) of a type with some *Adjective-Cluster* prefix.

Example 5.
```
definition
  cluster cardinal set;
    existence
      proof ...
      end;
end;
```

in article card_1 asserts the existence of the type "set" with the *Adjective* "cardinal". Whenever a variable is being reserved for a type with some non empty *Adjective-Cluster*, the non emptiness of the type's extension is checked by the system. We already noted how types with *Adjective-Cluster* are translated, so the translation is a simple existence formula in such cases:
"exists([X], (cardinal(X) and set(X)))".
In absolute notation and pruning the "set":
"exists([X], v1_card_1(X))".

Conditional-Registration. The syntax of *Conditional-Registration* is:
cluster *Adjective-Cluster* -> *Adjective-Cluster Type-Expression* ;
Correctness-Conditions .

Conditional-Registrations give rules for attribute propagation, e.g.:

Example 6.
```
definition
  cluster cardinal -> Ordinal-like set;
    coherence
      proof ...
      end;
end;
```

from article card_1 states that to any "set" with *Adjective-Cluster* "cardinal" the system should add the *Adjective-Cluster* "Ordinal-like". A proof of this must be given after the keyword "coherence". *Conditional-Registrations* are easily translated as implication:

"set(X) and cardinal(X) implies Ordinal-like(X)".
In absolute syntax and after "set" pruning:
"forall([X], implies(v1_card_1(X), v3_ordinal1(X)))".

Functorial-Registration. The syntax of *Functorial-Registration* is:
cluster *Term-Expression* -> *Adjective-Cluster* ; *Correctness-Conditions*

Example 7.
```
definition let X be finite set;
 cluster Card X -> finite;
 coherence;
end;
```
from article card_1 says that for any X with the type "finite set", the attribute "finite" applies also to the term "Card X" (the cardinality of X). While *Conditional-Registration*s define attribute propagation for some underlying type ("set" in the previous example), *Functorial-Registration*s do this for terms ("Card X" here). Their translation is also simple:
"(finite(X) and set(x)) implies finite(Card(X))".
In absolute syntax and after "set" pruning :
"(forall([X], implies(v1_finset_1(X), v1_finset_1(k1_card_1(X)))))".

4 Redefinitions

In Mizar, there can be predicate, attribute, type and functor redefinitions. Apart from the keyword "redefine", they have the same syntactic structure as the respective definitions. They can introduce new notation, change the definitional formula or the result types, or add new properties. Some redefinitions do not touch the constructor level, they only define new notation.

If the constructor level is touched by the redefinition, Mizar usually creates a new constructor, but knows to some extent, that it is equivalent to its redefined counterpart. The treatment in Mizar is not complete, e.g. if several type redefinitions are possible, only the last defined is chosen, since Mizar implementation sometimes requires the terms to have unique result types. This is no serious restriction for Mizar, since Mizar allows explicit typing (using the "is" keyword) of terms.

Since we translate types (and attributes) as predicates, there is no trouble in having multiple "types" for a term. After the translation, these are simply several unit clauses holding about the term, e.g.:
"finite(A) and Ordinal(A) and Integer(A) and Real(A)".
So unlike Mizar, we do not create new constructors for redefinitions, we just translate the new facts stated by the redefinitions.

5 Unimplemented Features of Mizar

Mizar *Scheme*s and Fraenkel terms are not currently translated, numbers are translated only in a very simple way.

Mizar *Schemes* are used to express second-order assertions parameterized by functor or predicate variables. This makes it possible, to state principles like Separation or Induction in Mizar, e.g.:

Example 8. `scheme Separation { A()-> set, P[set] }` :
  `ex X st for x holds x in X iff x in A() & P[x]`
  `proof ...`
  `end;`

from article `boole` states that for any set "`A()`" and any formula "`P`" with a free variable of the type "`set`", there is a subset of "`A()`" of elements with the property "`P`".

Another feature of Mizar is the Fraenkel ("setof") operator, which (in accordance with the Separation principle) creates sets of elements that satisfy a given formula.

Example 9. `{i where i is Nat: i < 5}`

is the set of natural numbers smaller than 5. The result type of such terms is the largest type "`set`".

Schemes and Fraenkel terms cannot be directly translated into first-order syntax. These features are relatively rare in MML and their usage is quite restricted (full instantiation of a scheme must be given before it can be applied), so a very large part of MML can be translated even if these features stay unhandled. We think the performance gap between current first order provers and higher order provers justifies this omission.

Theoretically, the Fraenkel terms and *Schemes* could be completely removed using the standard set-theoretic elimination procedures (previous work with Otter in this direction is described in [Belinfante 96]), however, given the advanced constructor system in Mizar, this could lead to very large (and thus practically unusable) expansions. We believe, that the solution lies rather in modifying the current efficient first order provers to handle the "real mathematics" (at least to the extent the Mizar checker handles it today). This also applies to the Mizar built-in handling of numbers. The current (most simple) translation creates for each mentioned number a new constant. However, Mizar can evaluate some simple arithmetic expressions like "`(5+3)*4`".

6   Related and Future Work

An important part of the translation are the scripts that generate proving problems from the translated Mizar database. Such scripts have to mimic the Mizar theory inclusion mechanism closely, add special axioms for things that are obvious to the Mizar checker, and possibly try to apply some filtering to handle large theories. Due to space constraints, we do not describe them here, they will probably be described in another article about experiments with theorem provers on the translated database.

Experimenting with theorem provers (currently E and SPASS) is work in progress at the time of writing this article. The main purpose behind the translation is to train theorem provers on the large database, and possibly also implement discovery systems that could make use of it.

The most recent experiments show, that the "well-typing" inferences that have to be carried out by the provers are much less efficient than the type system implementation in Mizar. So another work which is currently under way, is to add an efficient Mizar-like type system to theorem provers. It seems that implementation of type discipline will be a necessary feature for theorem provers to be able to handle large theories. For example, the SPASS prover, which implements Mizar types to some extent, seems to perform much better in advanced theories, than the heavily optimized recent CASC winners.

As already noted, another feature, that will eventually have to be added to theorem provers to handle theories based on ZFC (i.e. most of current mathematics), is efficient implementation of infinite schemes of first-order axioms. We also believe (cf. [Schumann 2001]), that provers should handle numbers efficiently.

We think a good way to boost the development of all these features for current theorem provers, would be to have e.g. a special part of CASC devoted to proving in large structured theory ensembles like MML. Detailed rules of such competition concerning e.g. allowed machine learning or type translation methods would have to be thought up.

References

[Belinfante 96] Johan Belinfante, On a Modification of Godel's Algorithm for Class Formation, Association for Automated Reasoning Newsletter, No. 34, pp. 1015 (1996)]

[Bonarska 90] Bonarska, E., An Introduction to PC Mizar, Fondation Ph. le Hodey, Brussels, 1990.

[Dahn 97] First Order Proof Problems Extracted from an Article in the MIZAR Mathematical Library. Ingo Dahn and Christoph Wernhard Proceedings of the International Workshop on First order Theorem Proving. RISC-Linz Report Series, No. 97–50, Johannes Kepler Universität Linz, 1997.

[Dahn 98] Ingo Dahn. Interpretation of a Mizar-like Logic in First Order Logic. Proceedings of FTP 1998. pp. 137–151.

[Goguen 92] J.Goguen and J. Meseguer. Order-sorted algebra I: Equational deduction formultiple inheritance, overloading, exceptions and partial operations. Theoretical Computer Science, 105(2):217–273,1992.

[Hahnle et all 96] R. Hahnle, M. Kerber, and C. Weidenbach. Common Syntax of the DFGSchwerpunktprogramm Deduction. Technical Report TR 10/96, Fakultät für Informatik, Universät Karlsruhe, Karlsruhe, Germany, 1996.

[McCune 94] McCune, W. W., OTTER 3.0 Reference Manual and Guide, Technical Report ANL-94/6, Argonne National Laboratory, Argonne, Illinois (1994).

[Muzalewski 93] Muzalewski, M., An Outline of PC Mizar, Fondation Philippe le Hodey, Brussels, 1993.

[REAL_1] Krzysztof Hryniewiecki, Basic properties of real numbers. Journal of Formalized Mathematics, 1, 1989.

[Rudnicky 92] Rudnicki, P., An Overview of the Mizar Project, Proceedings of the 1992 Workshop on Types for Proofs and Programs, Chalmers University of Technology, Bastad, 1992.

[Schumann 2001] J. M. Schumann, Automated Theorem-Proving in Software Engineering. Springer-Verlag, 2001.

[Suttner and Sutcliffe 98] C. Suttner and G. Sutcliffe. The TPTP problem library (TPTP v2.2.0). Technical Report 9704, Department of Computer Science, James Cook University, Townsville, Australia, 1998.

[Syntax] Syntax of the Mizar Language available online at
http://mizar.uwb.edu.pl/language/syntax.html

[Weidenbach et all 99] Weidenbach C., Afshordel B., Brahm U., Cohrs C., Engel T., Keen R., Theobalt C. and Topic D., System description: Spass version 1.0.0, in H. Ganzinger, ed., '16th International Conference on Automated Deduction, CADE-16', Vol. 1632 of LNAI, Springer, pp 314–318

[Wiedijk 99] Freek Wiedijk. Mizar: An impression. Unpublished paper, 1999.
http://www.cs.kun.nl/~freek/mizar/mizarintro.ps.gz.

The Mathematical Semantic Web

Massimo Marchiori

[1] The World Wide Web Consortium, MIT, Cambridge, USA
[2] Department of Computer Science, University of Venice, Italy
massimo@w3.org

Abstract. How can Mathematics and the Semantic Web effectively join? In this paper we provide an account of the key standard technologies that can foster the integration of Mathematical representation into the Semantic Web.

1 Introduction

Modern approaches to Mathematics on the web, nowadays, have in common with XML the well-known dualism between semantics and presentation: the media should be detached from the meaning ([1]). With the advent of the Semantic Web, however, the potential of this dualism seems to raise even higher, as that "semantics" part can possibly be more than itself, be part of a whole, be a single brick in a big building, the World Wide Web. For Math, this means that not only the Semantic Structure and Presentation Structure play their obvious classical role, but new possibilities come from the integration of various mathematical sources, allowing more powerful global search capabilities, but also associated metadata (context), possibility of computable math (algebraic manipulation / calculation) and so on. In order to achieve this, however, there is the need for integration: integration of the mathematical knowledge expertize/techniques, with the mainstream technologies that constitute the Semantic Web. Here, we focus on those Web technologies that could be potentially low-hanging fruits: RDF, OWL, XML-Schema, XML-Query and Functions and Operators.

2 RDF

When one talks about the Semantic Web, the first thing that comes to mind is RDF, the milestone on which the classic Semantic Web architecture lies. RDF, or Resource Description Framework (cf. [9,8,3]), is essentially an enriched entity-relationship model, designed to encode data (and, so to say, "meaning") in the World Wide Web.

So, while there are already XML representations for Mathematics on the Web (the most prominent example being MathML), in order to fully enter the Semantic Web bandwagon, there is the need to go beyond a normal XML representation, and try instead to link a bridge towards RDF-land.

A. Asperti, B. Buchberger, J.H. Davenport (Eds.): MKM 2003, LNCS 2594, pp. 216–223, 2003.
© Springer-Verlag Berlin Heidelberg 2003

This doesn't mean that existing XML dialects for Mathematics have to be abandoned, or that future versions and improvements have to be based on RDF. Rather, it means that a suitable representation to and from RDF should be given, so that applications dealing with Mathematical knowledge can export their knowledge to the Semantic Web (using an RDF representation), and conversely, that such an RDF representation can be mapped back[1] An example of such approach is for example given by the P3P standard ([11]), which has its own XML syntax, and a corresponding alternate RDF representation ([12]).

Many straightforward representations of Math knowledge to RDF are possible, given that RDF expresses, as said, a generic graph structure (entity-relationship) that can be made to fit any generic structure. In particular, it is rather evident that mathematical formulas, for their same nature of being usually serializable in text form, fit rather well in the RDF graph model. For example, the representation of a function $p(2, 6)$ expressed in MathML as

```
<apply>
  <csymbol encoding="text"
           definitionURL="http://www.mathsw.org/scalarplus">
           p
           </csymbol>
  <cn> 2 </cn>
  <cn> 6 </cn>
</apply>
```

could be given an RDF representation like:

```
:_1 <http://www.w3.org/TR/MathML2#apply> :_2
:_1 <http://www.w3.org/TR/MathML2#csymbol> "p"
:_1 <http://www.w3.org/TR/MathML2#definitionURL>
                              <http://www.mathsw.org/scalarplus>
:_1 <http://www.w3.org/TR/MathML2#encoding> "text"
:_2 <http://www.w3.org/1999/02/22-rdf-syntax-ns#:_1> "2"
:_2 <http://www.w3.org/1999/02/22-rdf-syntax-ns#:_2> "6"
```

Here, we have used the so-called "triple" representation of an RDF graph, where each row represents a triple

entity1 relationship entity2

(see e.g. [8] for more details).

Much like *proof nets*, the RDF graph structure nicely leads to encoding mathematical structures (it allows for example easy (and effective) processing of bound variables). For example, the function $\lambda x. \; x + x$ could be encoded in MathML with an explicit bound variable like this:

---

[1] Although, what is really needed is the mapping from the XML dialect to RDF, as the reverse mapping could be (depending on the specific application) of less use, or even impossible in some cases (when the domain of discourse is enlarged in RDF).

```
<lambda>
  <bvar><ci> x </ci></bvar>
    <apply>
      <plus/>
      <ci> x </ci>
      <ci> x </ci>
    </apply>
```

RDF, on the other hand, allows a neater representation of the bound variable:

```
:_1 <http://www.w3.org/TR/MathML2#apply> :_3
:_1 <http://www.w3.org/TR/MathML2#lambda> :_2
:_3 <http://www.w3.org/TR/MathML2#definitionURL>
                          <http://www.w3.org/TR/MathML2#plus>
:_3 <http://www.w3.org/1999/02/22-rdf-syntax-ns#:_1> :_2
:_3 <http://www.w3.org/1999/02/22-rdf-syntax-ns#:_2> :_2
```

The technical details of any such mapping to RDF are unnecessary here, as first they are rather obvious[2], and second, many equivalent choices are possible. What is important, though, is to note that the mathematical structure of objects can/should be given an RDF formulation, so to foster integration between the Math world and the Semantic Web. With an RDF representation for Math, we can use all the tools developed/in development for the Semantic Web, like for example search, inference, annotation, and so on. So, general-purpose tools can be reused to operate with mathematical objects, and no special-purpose tools have to be developed. Most importantly, putting mathematics inside RDF allows a nice and smooth integration with all various other sources of data: in other words, proper integration with the Semantic Web via RDF allows for a seamless definition of *contexts*, and linkage to various other metadata related to the math object. This allows to go way beyond the usual way of allowing metadata in XML dialects, i.e., using specific attributes where to fit more information. So, RDF really allows making Mathematics integrated in the World Wide Web.

3   OWL

Mathematical objects "per se" can be very difficult objects to categorize, if appropriate metadata is not provided. This is a general problem of many kinds of data present on the Web, so much that W3C has addressed this issue by chartering a new effort for the production of a standard devoted to Ontologies for the Web: the Web Ontology Language (also called "OWL"), see for instance [7]. OWL allows data to be attached with ontological meaning: in other words, it allows powerful mechanisms to categorize data on the web into specific classes and subclasses, therefore greatly facilitating the proper handling/search/query of web data.

---

[2] Although, note that even from the simple examples shown here, some subtleties are present, and we have been for example sloppy in the treatment of datatypes, as we will hint at later.

What this means from a Math viewpoint is that mathematical objects, that can by their nature be rather dense and not easily categorizable at a first inspection, can instead be nicely associated to some specific categories. Such mathematical ontologies can provide a formidable help for the task of searching for mathematics, and for the task of issuing similarity searches, avoiding the intrinsic difficulties of extracting this information from the semantics/structure of the object (which might obviously be just plain impossible). OWL is flexible enough to allow for complete mathematical hierarchies to be formed, and also to allow for alternate hierarchies, so that no just one categorization has to be used, but many can happily coexist.

4   XML Schema

Types are an essential component in mathematics, and it comes by no surprise that even when attempting to formalize mathematics on the Web, types have entered the battlefield. Numeric types like integer, real, complex and so on have been defined in MathML, while the OpenMath Consortium has also experimentally tried to give more complex typing facilities using the Extended Calculus of Constructors ([6]). On the other hand, type systems have also been introduced in the Web, and more specifically, one type system is now the world standard: XML Schema ([13,4]). It is tempting therefore to analyze whether there are differences, and where convergence is possible here.

The first big distinction that has to be done is between what XML-Schema calls *primitive* datatypes. In a sense, primitive types are the basic building blocks, from which more complex types can then be created. XML-Schema primitive datatypes much correspond, in spirit, to MathML types. However, we can see that there are differences, as MathML focusses on numeric types, and neglects other specific XML datatypes. Now, here convergency is not only possible (primitive datatypes can be merged with no problems), but strongly needed for the future integration of mathematics on the web. In fact, RDF is already adopting XML-Schema primitive datatypes in the Semantic Web view of the world, which makes even more urgent for mathematical knowledge to align. Note that numeric datatypes currently not present in XML Schema can be added using so-called user defined types, so the merge of the two needs (XML-Schema primitive types, and "primitive types" of the mathematical community) can fit within XML-Schema.

The other part of the XML-Schema type system is more powerful, but here, the two types of schemas (XML, and math) diverge in a sense, as the first one deals with structural constraints, while the second deals more with signature definition. It seems like XML-Schema type system could accommodate some math type systems, but the interaction here is deeper, as some desired type system for math (like the ECC mentioned above) are more fine-grained than XML-Schema's type system in some circumstances, so convergence here should be explored further.

5   Functions and Operators

Types are useful for a variety of task, last but not least, validation. But another very useful functionality they provide is just to provide some signatures, so that *executable specification* of the corresponding function/operator can be safely activated, even when no full or accurate type system is present. And one of the possible cool functionalities that math functions an operators could have is just this: to be *computable*, in the sense that they could, somehow, be activated (evaluated), and the appropriate result returned. There are of course various model that would allow to have executable mathematics on the Web, from the central server model (where requests are sent, for example using a Web Service architecture, or classic HTTP PUT/POST methods), to the local server model, where applets could be for example downloaded, and executed in situ. What matters most for our discussion is, besides the need for a uniform way (or, uniform multiple ways) to perform such activation, the necessity to have uniformity.

The first key to successfully represent mathematics on the web is a single acronym: URIs. As Web Architecture dictates, URIs should be used to denote every relevant thing in the Web. Restated, in the mathematical environment every function/operators/relations and so on, should be given an appropriate URI, that is to say, an appropriate name in Web space. It is for this reason that the current MathML standard ([5]) allows for new symbols to be classified using a URI: this allows for much better search on the web, as well as, for example, the possibility of automatic execution and/or algebraic manipulation between different sources of math. Without URIs, we have to stick to local names, and therefore we have names that are useful only within a single document, or within a single application.

But, if URIs provides us with good "global addresses" along the Web, the other crucial factor for their success is that, as far as possible, people don't introduce new URIs that always denote the same thing. If this happens, the utility of the URI boils down to just a single identifier, making the web just an isolated bunch of points, with no connections.

In our case, this means that it is of little use to use URIs for functions and operators, if each of us creates such URIs from scratch, without reusing existing URIs. There is the need for some appropriate standardization of common operators, so that people can reuse it, search tools can link different occurrences in different parts of the web, execution of a vast number of functions can be performed. [10] is an example of a first catalog of functions and operators that are going to be standardized, for uniform usage over the Web. Similar efforts could be undertaken to extend such collection to more mathematical functions, where first-order functions and operators could be considered first. In any case, it is obviously important that uniformity prevails, and that there is convergence between the XML/database world, and the mathematical environment on the web (think for example at the collection in [14]).

6 XML Query

From what seen previously, it would seem like RDF and OWL already provide all what we need for a proper integration of Mathematical knowledge into the Semantic Web. But of course, the complete picture is far from being that simple. While, certainly, RDF and OWL can do a lot to foster reuse of Semantic Web tools with mathematical objects, there are still some drawbacks that one has to take into proper account.

The primary of such drawback is the complexity of the tools. Simply having an RDF representation doesn't solve our needs, as then we need to use some tools that allow us to query and manipulate the mathematical objects. Therefore, another aspect has to be considered: the tradeoff between the complexity of such tools, and our needs.

As far as the functional requirements are concerned, we envisage that, to start with, the primary applications in the use of mathematical knowledge on the World Wide Web in the large will be forms of *simple search*, possibly *extraction*, and easy *manipulation*.

Moreover, always talking in the large, the other factor to consider here is the scalability of the solution. The RDF representation make mathematical objects migrate to a more general space, the RDF graph, at a price: it can complicates quite a lot the structure, as compared to the original XML representation. When talking about search/extraction/manipulation in the large, therefore, we risk that staying in the RDF graph for such operations will not scale so easily, leading to *failure of responsiveness*.

The possible way out to this is not to give away the XML representation when we need it, as in this case. Such representation can in fact be traversed much more efficiently than an RDF representation, and as such, specific tools that operate on XML can be used. In particular, XML Query (also known as XQuery, cf. [2]), the future world standard for querying XML data, could be profitably reused. XML Query itself operates on a variant of the relational model, adapted to the specific XML data model, and has been designed with all the needs that database vendors seek out in current SQL-like systems,, like, in primis, speed and scalability. It is an interesting exercise to see, given a certain XML representation for math (like, say, MathML), what is the power of queries and related operations that XQuery can perform[3] . For sure, XQuery fits perfectly in the use cases of simple search and extraction, which makes it a good candidate for query/search application of XML Mathematical knowledge in the Web (so, in the large). In fact, for more sophisticated applications, one could imagine the possibility of a *hybrid* system, where the mathematical objects are just seen in the RDF graph as XML literals. This still allows the mathematical objects to interact with the semantic web, to be given context, annotations and so on. What we would lose is easy access to the internal structure of the objects, within the RDF world. For that access, we could activate an XQuery processor, that could quickly find the result we need. And in fact, such XML literals could always be exploded "on demand" to their RDF

---

[3] Indeed, XQuery, although *Turing complete*, is not *functionally complete*.

structure, like in lazy functional programming, when really needed. Therefore, the hybrid model would work well in all those situation where full RDF access to the internal structure of the mathematical objects is seldom needed.

7 Success?

Having seen the possible synergies, the final questions to ask is: can this integration be performed with success?

The first factor to consider is merely technical, and has to do with the problem of *vastity of domain.*
Mathematics is a vast field. This means that potentially, Semantic Web application like search, similarity search, manipulation, inference, can perform badly if not properly assisted with precise Semantic Web annotations. There is probably the need to identify some critical subsets of mathematical knowledge that can benefit more from semantical structuring (pretty much the same kind of selection that MathML somehow had to do when facing the creation of an XML math dialect); this seems even more important when we go to the higher layers (computability, typing, algebraic manipulation). Constructive mathematics here obviously will play a much more relevant role.

The second factor to consider is societal, and concerns the *cost/benefit vicious circle.*
Information encoded using the semantic web has a higher cost, like any form of evolved semantic encoding, than a merely syntactic formatting. So, the cost to put information on the Web gets higher. In order to make people accept this extra cost, the cost/ratio benefit must stay low. But crucially, in the semantic web the benefits usually depends on adoption: the more, the better. Therefore, we have this vicious circle: people will not be very likely to contribute to a mathematical semantic web, unless critical mass is reached. It is this vicious circle that might slow down too much progress in the field, if not carefully weighted.

References

1. A.Asperti, L.Padovani, C.S.Coen, I.Schena, *XML, Stylesheets and the rem-mathematization of Formal Content*, Proceedings of "Extreme Markup Languages 2001 Conference", 2001. Available at
   http://www.cs.unibo.it/helm/extreme2001.ps.gz.
2. S.Boag *et al.* (Eds.), *XQuery 1.0: An XML Query Language*, World Wide Web Consortium Working Draft, November 2002. Latest version at
   http://www.w3.org/TR/xquery.
3. D.Beckett (Ed.), *RDF/XML Syntax Specification (Revised)*, World Wide Web Consortium Working Draft, November 2002. Latest version at
   http://www.w3.org/TR/rdf-syntax-grammar.
4. P.V.Biron and A.Malhotra (Eds.), *XML Schema Part 2: Datatypes*, World Wide Web Consortium Recommendation, 2001. Available at
   http://www.w3.org/TR/xmlschema-2/.

5. D.Carlisle, P.Ion, R.Milner, N.Poppelier, *Mathematical Markup Language (MathML) Version 2.0*, World Wide Web Consortium Recommendation, 2001. Available at http://www.w3.org/TR/MathML2.

6. O.Caprotti and A.M.Cohen, *A Type System for OpenMath*, The OpenMath Consortium, 1999. Available at
http://monet.nag.co.uk/cocoon/openmath/standard/ecc.pdf.

7. M.Dean *et al* (Eds.), *Web Ontology Language (OWL) Reference Version 1.0*. World Wide Web Consortium Working Draft, November 2002. Latest version at http://www.w3.org/TR/owl-ref.

8. P.Hayes (Ed.), *RDF Semantics*, World Wide Web Consortium Working Draft, November 2002. Latest version at http://www.w3.org/TR/rdf-mt/.

9. O.Lassila and R.Swick (Eds.), *Resource Description Framework (RDF) Model and Syntax Specification*, World Wide Web Consortium Recommendation, 1999. Available at http://www.w3.org/TR/REC-rdf-syntax.

10. A.Malhotra *et al.* (Eds.), *XQuery 1.0 and XPath 2.0 Functions and Operators*, World Wide Web Consortium Working Draft, November 2002. Latest version at http://www.w3.org/TR/xquery-operators/.

11. M.Marchiori (Ed.), *The Platform for Privacy Preferences 1.0 (P3P1.0) Specification*, World Wide Web Consortium
Recommendation, 2002. Available at http://www.w3.org/TR/P3P.

12. B.McBride *et al.*, *An RDF Schema for P3P*, World Wide Web Consortium Note. Latest version at http://www.w3.org/TR/p3p-rdfschema/.

13. H.Thompson *et al.* (Eds.), *XML Schema Part 1: Structures*, World Wide Web Consortium Recommendation, 2001. Available at
http://www.w3.org/TR/xmlschema-1/.

14. Wolfram Research, *Mathematical Functions*. Available at http://functions.wolfram.com/.

Author Index

Adams, Andrew A.   1
Anghelache, Romeo   147

Baba, Yusuke   93
Bancerek, Grzegorz   119
Borwein, Jonathan   45

Cairns, Paul   80, 175
Carlisle, David   56
Coen, Claudio Sacerdoti   30

Davenport, James H.   17
Dewar, Mike   56

Goguadze, Georgi   80
Gow, Jeremy   175
Güntzer, Ulrich   133
Guidi, Ferruccio   105

Heumesser, Bernd D.   133

Kohlhase, Michael   147

Marchiori, Massimo   216
Melis, Erica   80

Padovani, Luca   66

Rudnicki, Piotr   119, 162

Schena, Irene   105
Seipel, Dietmar A.   133
Stanway, Terry   45
Suzuki, Masakazu   93

Trybulec, Andrzej   162

Ullrich, Carsten   80
Urban, Josef   203

Wiedijk, Freek   188

Lecture Notes in Computer Science

For information about Vols. 1–2493

please contact your bookseller or Springer-Verlag

Vol. 2494: B.W. Watson, D. Wood (Eds.), Implementation and Application of Automata. Proceedings, 2001. X, 289 pages. 2002.

Vol. 2495: C. George, H. Miao (Eds.), Formal Methods and Software Engineering. Proceedings, 2002. XI, 626 pages. 2002.

Vol. 2496: K.C. Almeroth, M. Hasan (Eds.), Management of Multimedia in the Internet. Proceedings, 2002. XI, 355 pages. 2002.

Vol. 2497: E. Gregori, G. Anastasi, S. Basagni (Eds.), Advanced Lectures on Networking. XI, 195 pages. 2002.

Vol. 2498: G. Borriello, L.E. Holmquist (Eds.), UbiComp 2002: Ubiquitous Computing. Proceedings, 2002. XV, 380 pages. 2002.

Vol. 2499: S.D. Richardson (Ed.), Machine Translation: From Research to Real Users. Proceedings, 2002. XXI, 254 pages. 2002. (Subseries LNAI).

Vol. 2500: E. Grädel, W. Thomas, T. Wilke (Eds.), Automata Logics, and Infinite Games. VIII, 385 pages. 2002.

Vol. 2501: D. Zheng (Ed.), Advances in Cryptology – ASIACRYPT 2002. Proceedings, 2002. XIII, 578 pages. 2002.

Vol. 2502: D. Gollmann, G. Karjoth, M. Waidner (Eds.), Computer Security – ESORICS 2002. Proceedings, 2002. X, 281 pages. 2002.

Vol. 2503: S. Spaccapietra, S.T. March, Y. Kambayashi (Eds.), Conceptual Modeling – ER 2002. Proceedings, 2002. XX, 480 pages. 2002.

Vol. 2504: M.T. Escrig, F. Toledo, E. Golobardes (Eds.), Topics in Artificial Intelligence. Proceedings, 2002. XI, 432 pages. 2002. (Subseries LNAI).

Vol. 2506: M. Feridun, P. Kropf, G. Babin (Eds.), Management Technologies for E-Commerce and E-Business Applications. Proceedings, 2002. IX, 209 pages. 2002.

Vol. 2507: G. Bittencourt, G.L. Ramalho (Eds.), Advances in Artificial Intelligence. Proceedings, 2002. XIII, 418 pages. 2002. (Subseries LNAI).

Vol. 2508: D. Malkhi (Ed.), Distributed Computing. Proceedings, 2002. X, 371 pages. 2002.

Vol. 2509: C.S. Calude, M.J. Dinneen, F. Peper (Eds.), Unconventional Models in Computation. Proceedings, 2002. VIII, 331 pages. 2002.

Vol. 2510: H. Shafazand, A Min Tjoa (Eds.), EurAsia-ICT 2002: Information and Communication Technology. Proceedings, 2002. XXIII, 1020 pages. 2002.

Vol. 2511: B. Stiller, M. Smirnow, M. Karsten, P. Reichl (Eds.), From QoS Provisioning to QoS Charging. Proceedings, 2002. XIV, 348 pages. 2002.

Vol. 2512: C. Bussler, R. Hull, S. McIlraith, M.E. Orlowska, B. Pernici, J. Yang (Eds.), Web Services, E-Business, and the Semantic Web. Proceedings, 2002. XI, 277 pages. 2002.

Vol. 2513: R. Deng, S. Qing, F. Bao, J. Zhou (Eds.), Information and Communications Security. Proceedings, 2002. XII, 496 pages. 2002.

Vol. 2514: M. Baaz, A. Voronkov (Eds.), Logic for Programming, Artificial Intelligence, and Reasoning. Proceedings, 2002. XIII, 465 pages. 2002. (Subseries LNAI).

Vol. 2515: F. Boavida, E. Monteiro, J. Orvalho (Eds.), Protocols and Systems for Interactive Distributed Multimedia. Proceedings, 2002. XIV, 372 pages. 2002.

Vol. 2516: A. Wespi, G. Vigna, L. Deri (Eds.), Recent Advances in Intrusion Detection. Proceedings, 2002. X, 327 pages. 2002.

Vol. 2517: M.D. Aagaard, J.W. O'Leary (Eds.), Formal Methods in Computer-Aided Design. Proceedings, 2002. XI, 399 pages. 2002.

Vol. 2518: P. Bose, P. Morin (Eds.), Algorithms and Computation. Proceedings, 2002. XIII, 656 pages. 2002.

Vol. 2519: R. Meersman, Z. Tari, et al. (Eds.), On the Move to Meaningful Internet Systems 2002: CoopIS, DOA, and ODBASE. Proceedings, 2002. XXIII, 1367 pages. 2002.

Vol. 2521: A. Karmouch, T. Magedanz, J. Delgado (Eds.), Mobile Agents for Telecommunication Applications. Proceedings, 2002. XII, 317 pages. 2002.

Vol. 2522: T. Andreasen, A. Motro, H. Christiansen, H. Legind Larsen (Eds.), Flexible Query Answering. Proceedings, 2002. XI, 386 pages. 2002. (Subseries LNAI).

Vol. 2523: B.S. Kaliski Jr., Ç.K. Koç, C. Paar (Eds.), Cryptographic Hardware and Embedded Systems – CHES 2002. Proceedings, 2002. XIV, 612 pages. 2002.

Vol. 2525: H.H. Bülthoff, S.-Whan Lee, T.A. Poggio, C. Wallraven (Eds.), Biologically Motivated Computer Vision. Proceedings, 2002. XIV, 662 pages. 2002.

Vol. 2526: A. Colosimo, A. Giuliani, P. Sirabella (Eds.), Medical Data Analysis. Proceedings, 2002. IX, 222 pages. 2002.

Vol. 2527: F.J. Garijo, J.C. Riquelme, M. Toro (Eds.), Advances in Artificial Intelligence – IBERAMIA 2002. Proceedings, 2002. XVIII, 955 pages. 2002. (Subseries LNAI).

Vol. 2528: M.T. Goodrich, S.G. Kobourov (Eds.), Graph Drawing. Proceedings, 2002. XIII, 384 pages. 2002.

Vol. 2529: D.A. Peled, M.Y. Vardi (Eds.), Formal Techniques for Networked and Distributed Sytems – FORTE 2002. Proceedings, 2002. XI, 371 pages. 2002.

Vol. 2531: J. Padget, O. Shehory, D. Parkes, N. Sadeh, W.E. Walsh (Eds.), Agent-Mediated Electronic Commerce IV. Proceedings, 2002. XVII, 341 pages. 2002. (Subseries LNAI).

Vol. 2532: Y.-C. Chen, L.-W. Chang, C.-T. Hsu (Eds.), Advances in Multimedia Information Processing – PCM 2002. Proceedings, 2002. XXI, 1255 pages. 2002.

Vol. 2533: N. Cesa-Bianchi, M. Numao, R. Reischuk (Eds.), Algorithmic Learning Theory. Proceedings, 2002. XI, 415 pages. 2002. (Subseries LNAI).

Vol. 2534: S. Lange, K. Satoh, C.H. Smith (Ed.), Discovery Science. Proceedings, 2002. XIII, 464 pages. 2002.

Vol. 2535: N. Suri (Ed.), Mobile Agents. Proceedings, 2002. X, 203 pages. 2002.

Vol. 2536: M. Parashar (Ed.), Grid Computing – GRID 2002. Proceedings, 2002. XI, 318 pages. 2002.

Vol. 2537: D.G. Feitelson, L. Rudolph, U. Schwiegelshohn (Eds.), Job Scheduling Strategies for Parallel Processing. Proceedings, 2002. VII, 237 pages. 2002.

Vol. 2538: B. König-Ries, K. Makki, S.A.M. Makki, N. Pissinou, P. Scheuermann (Eds.), Developing an Infrastructure for Mobile and Wireless Systems. Proceedings 2001. X, 183 pages. 2002.

Vol. 2539: K. Börner, C. Chen (Eds.), Visual Interfaces to Digital Libraries. X, 233 pages. 2002.

Vol. 2540: W.I. Grosky, F. Plášil (Eds.), SOFSEM 2002: Theory and Practice of Informatics. Proceedings, 2002. X, 289 pages. 2002.

Vol. 2541: T. Barkowsky, Mental Representation and Processing of Geographic Knowledge. X, 174 pages. 2002. (Subseries LNAI).

Vol. 2544: S. Bhalla (Ed.), Databases in Networked Information Systems. Proceedings 2002. X, 285 pages. 2002.

Vol. 2545: P. Forbrig, Q, Limbourg, B. Urban, J. Vanderdonckt (Eds.), Interactive Systems. Proceedings 2002. X, 269 pages. 2002.

Vol. 2546: J. Sterbenz, O. Takada, C. Tschudin, B. Plattner (Eds.), Active Networks. Proceedings, 2002. XIV, 267 pages. 2002.

Vol. 2547: R. Fleischer, B. Moret, E. Meineche Schmidt (Eds.), Experimental Algorithmics. XVII, 279 pages. 2002.

Vol. 2548: J. Hernández, Ana Moreira (Eds.), Object-Oriented Technology. Proceedings, 2002. VIII, 223 pages. 2002.

Vol. 2549: J. Cortadella, A. Yakovlev, G. Rozenberg (Eds.), Concurrency and Hardware Design. XI, 345 pages. 2002.

Vol. 2550: A. Jean-Marie (Ed.), Advances in Computing Science – ASIAN 2002. Proceedings, 2002. X, 233 pages. 2002.

Vol. 2551: A. Menezes, P. Sarkar (Eds.), Progress in Cryptology – INDOCRYPT 2002. Proceedings, 2002. XI, 437 pages. 2002.

Vol. 2552: S. Sahni, V.K. Prasanna, U. Shukla (Eds.), High Performance Computing – HiPC 2002. Proceedings, 2002. XXI, 735 pages. 2002.

Vol. 2553: B. Andersson, M. Bergholtz, P. Johannesson (Eds.), Natural Language Processing and Information Systems. Proceedings, 2002. X, 241 pages. 2002.

Vol. 2554: M. Beetz, Plan-Based Control of Robotic Agents. XI, 191 pages. 2002. (Subseries LNAI).

Vol. 2555: E.-P. Lim, S. Foo, C. Khoo, H. Chen, E. Fox, S. Urs, T. Costantino (Eds.), Digital Libraries: People, Knowledge, and Technology. Proceedings, 2002. XVII, 535 pages. 2002.

Vol. 2556: M. Agrawal, A. Seth (Eds.), FST TCS 2002: Foundations of Software Technology and Theoretical Computer Science. Proceedings, 2002. XI, 361 pages. 2002.

Vol. 2557: B. McKay, J. Slaney (Eds.), AI 2002: Advances in Artificial Intelligence. Proceedings, 2002. XV, 730 pages. 2002. (Subseries LNAI).

Vol. 2558: P. Perner, Data Mining on Multimedia Data. X, 131 pages. 2002.

Vol. 2559: M. Oivo, S. Komi-Sirviö (Eds.), Product Focused Software Process Improvement. Proceedings, 2002. XV, 646 pages. 2002.

Vol. 2560: S. Goronzy, Robust Adaptation to Non-Native Accents in Automatic Speech Recognition. Proceedings, 2002. XI, 144 pages. 2002. (Subseries LNAI).

Vol. 2561: H.C.M. de Swart (Ed.), Relational Methods in Computer Science. Proceedings, 2001. X, 315 pages. 2002.

Vol. 2562: V. Dahl, P. Wadler (Eds.), Practical Aspects of Declarative Languages. Proceedings, 2003. X, 315 pages. 2002.

Vol. 2566: T.Æ. Mogensen, D.A. Schmidt, I.H. Sudborough (Eds.), The Essence of Computation. XIV, 473 pages. 2002.

Vol. 2567: Y.G. Desmedt (Ed.), Public Key Cryptography – PKC 2003. Proceedings, 2003. XI, 365 pages. 2002.

Vol. 2568: M. Hagiya, A. Ohuchi (Eds.), DNA Computing. Proceedings, 2002. XI, 338 pages. 2003.

Vol. 2569: D. Gollmann, G. Karjoth, M. Waidner (Eds.), Computer Security – ESORICS 2002. Proceedings, 2002. XIII, 648 pages. 2002. (Subseries LNAI).

Vol. 2571: S.K. Das, S. Bhattacharya (Eds.), Distributed Computing. Proceedings, 2002. XIV, 354 pages. 2002.

Vol. 2572: D. Calvanese, M. Lenzerini, R. Motwani (Eds.), Database Theory – ICDT 2003. Proceedings, 2003. XI, 455 pages. 2002.

Vol. 2574: M.-S. Chen, P.K. Chrysanthis, M. Sloman, A. Zaslavsky (Eds.), Mobile Data Management. Proceedings, 2003. XII, 414 pages. 2003.

Vol. 2575: L.D. Zuck, P.C. Attie, A. Cortesi, S. Mukhopadhyay (Eds.), Verification, Model Checking, and Abstract Interpretation. Proceedings, 2003. XI, 325 pages. 2003.

Vol. 2576: S. Cimato, C. Galdi, G. Persiano (Eds.), Security in Communication Networks. Proceedings, 2002. IX, 365 pages. 2003.

Vol. 2578: F.A.P. Petitcolas (Ed.), Information Hiding. Proceedings, 2002. IX, 427 pages. 2003.

Vol. 2580: H. Erdogmus, T. Weng (Eds.), COTS-Based Software Systems. Proceedings, 2003. XVIII, 261 pages. 2003.

Vol. 2588: A. Gelbukh (Ed.), Computational Linguistics and Intelligent Text Processing. Proceedings, 2003. XV, 648 pages. 2003.

Vol. 2594: A. Asperti, B. Buchberger, J.H. Davenport (Eds.), Mathematical Knowledge Management. Proceedings, 2003. X, 225 pages. 2003.

Vol. 2600: S. Mendelson, A.J. Smola, Advanced Lectures on Machine Learning. Proceedings, 2002. IX, 259 pages. 2003. (Subseries LNAI).